MATHEMATICS

A MINIMAL INTRODUCTION

MATHEMATICS

A MINIMAL INTRODUCTION

ALEXANDRU BUIUM

University of New Mexico
Albuquerque, USA

CRC Press
Taylor & Francis Group
Boca Raton   London   New York

CRC Press is an imprint of the
Taylor & Francis Group, an **informa** business

A CHAPMAN & HALL BOOK

CRC Press
Taylor & Francis Group
6000 Broken Sound Parkway NW, Suite 300
Boca Raton, FL 33487-2742

Printed on acid-free paper
Version Date: 20130916

International Standard Book Number-13: 978-1-4822-1600-4 (Paperback)

Visit the Taylor & Francis Web site at
http://www.taylorandfrancis.com

and the CRC Press Web site at
http://www.crcpress.com

Contents

Preface

This book is an undergraduate introduction to pure mathematics. It can serve as a course bridging the gap between "procedural" mathematics (that emphasizes calculation) and "conceptual" mathematics (that emphasizes ideas); it can also serve as a rigorous introduction to the basic concepts of logic and mathematics. This being said, the present book seems to drastically differ from most available textbooks, as we shall explain below.

Indeed most books introducing logic and mathematics start by assuming elementary mathematics as largely known in its "procedural" form; in particular they assume a certain familiarity with "handling" numbers, elementary geometric figures, and some elementary functions. This leads to a certain circularity in the presentation: mathematics is being used as an illustration (and even as a tool) in the construction of the logical apparatus while, on the other hand, logic is being used to construct mathematical proofs. This circularity is generally presented as acceptable, although, for excellent reasons, students are often uncomfortable accepting it. Moreover, most books on the subject implicitly subscribe to a correspondence theory of truth; according to such a theory mathematical sentences have a meaning, they are supposed to refer to an "infinite universe of mathematical reality," and they are supposed to be either true or false depending on their meaning and on the "state of affairs" in the mathematical universe. This leads to an implicit acceptance of a "maximal ontology," such as that of Cantor set theory, where "actual infinities" are being treated as "reality" and statements about them are being treated as referring to "empirical facts."

The present book will adopt, however, a radically different position: we will declare that the "infinite universe of mathematical reality" is a fiction; and that mathematics does not need this "fairy tale universe" in order to exist. All that mathematics needs are the mathematical words themselves which should be regulated by a "pre-mathematical logic" of language. For instance we will agree that empirical "infinity" is not available but that we can use the word "infinity" as a mere word, devoid of any reference, provided that strict rules are observed in its use. Consequently, in contrast with most available books, we will begin here by asking the students to "forget," for a while, about all the mathematical objects/concepts that they were ever exposed to; one will have to act as if one does not know the "meaning" of the words *implies, contradiction,* etc. or the "meaning" of the symbols 1, 2, 3, $+$, $\times$, $\mathbb{Z}$, $=$, sin, cos, etc. Then the course will introduce each of these words/symbols in a non-circular manner. In this way we will build, from scratch, first a *pre-mathematical logic*, then *mathematics* itself, and, finally, a *mathematical logic*. Pre-mathematical logic will deal exclusively with language and will be independent of mathematics; on the other hand mathematical logic will appear as a chapter of mathematics. Distinguishing between these two logics will help break the circularity mentioned earlier. On the other hand we will replace the "maximal ontology" of the "mathematical universe" with a "minimal ontology" according to which the only things "admitted into existence" will be (written) symbols and their combinations referred to as texts. Mathematical sentences will then be regarded as having no meaning and no reference; and everywhere in our proofs semantics will be replaced by syntax. A direct consequence of this will be that the concepts of truth and falsehood will become irrelevant in (and will be actually banned from) all of our discussion of mathematics.

The attitude of this book generally fits into what is called the *formalist* approach to mathematics, although the brand of formalism which we chose to adopt here is rather extreme. Nevertheless the author feels that extreme formalism can be easily grasped by students, who generally seem to enjoy playing the game of reconstructing mathematics from language itself. This is, of course, more than a game. Indeed one can argue that the only tenable alternatives to formalism seem to involve reduction of mathematics to either psychology (of the kind involved in intuitionism) or metaphysics (of the kind involved in platonism). Both these reductions seem to postulate a somewhat mysterious dimension beyond mathematics which this course would like to reject. There is also the pragmatic view that identifies mathematics with a toolbox for physical sciences. A moment's reflection shows that this approach is not as straightforward as it seems but, rather, runs into subtle epistemological problems; and attempts to solve these problems inevitably lead back to psychology or metaphysics. Finally there are other possible ways to introduce mathematics by viewing it not as discourse but as a creative process. One aspect of this, which, again, involves a mysterious psychological component, is the discovery of new mathematics by individual mathematicians. Another aspect is the evolution of mathematical ideas throughout history. By the way, there are quite a few excellent books written along historical lines; we kept historical comments to a minimum here because the historical order of things is often quite different from the order of things required by a non-circular mathematical discourse.

It should be clear by now why the present introduction to mathematics is called *minimal*. It is so called because it seeks to avoid any reference to non-mathematical disciplines such as metaphysics, physics, psychology, or history. What is left, after eliminating such references, is the mathematical discourse itself; and reaching this minimal point, where discourse is left to function by itself, is an intellectual adventure that should probably be part of any serious attempt to understand mathematics. Once this austere landscape of discourse has been revealed, students can start repopulating it, at will, with elements belonging to the various disciplines that we tried to avoid here.

The book is based upon courses taught in the Department of Mathematics and Statistics at the University of New Mexico, Albuquerque. I am especially indebted to my students in the Introduction to Mathematical Thinking course for their feedback and for lively interaction. I am also indebted to a number of mathematicians with whom, over the years, I had discussions on the subject of logic. Special thanks go to E. Bouscaren, P. Cartier, A. Pillay, and F. Pop.

Alexandru Buium
Albuquerque, May 2013

Introduction

In what follows we offer a general discussion of the problematic and contents of the course. This discussion below is not logically necessary for the understanding of the rest of the course and therefore may be skipped.

1. Mathematics versus logic. In this course we view mathematics as a chapter of logic. Logic organizes language into systems called theories. Then mathematics will be viewed as one among many theories; as a theory mathematics is generally viewed today as coinciding with set theory. Some of the key words of mathematics are: numbers, figures, and limits; they lead to chapters of mathematics called algebra, geometry, and analysis, respectively. This viewpoint on mathematics emerged from work of Frege, Cantor, Russell, and Hilbert at the turn of the 20th century; it evolved from a logicist to a formalist approach; this evolution roughly corresponds to the gradual elimination of semantics from set theory. The viewpoint that mathematics is, essentially, set theory was adopted by Bourbaki (a collective name for a group of leading French mathematicians in the middle of the 20th century), who provided such a presentation of mathematics in a series of now 9 classical monographs; in spite of various critical reactions to this project, today's (pure) mathematics is, essentially, Bourbaki mathematics. There are alternatives to set theory as foundation of mathematics; examples among recent ones are, for instance, Conway's construction of surreal numbers or the work of Voevodsky and others on the univalent foundations. We will not touch upon these (or other) alternatives here but, rather, stick to the set theoretic viewpoint which has dominated the 20-th century. Set theory itself has a number of variants; we chose here the Zermelo-Fraenkel+Choice variant which is probably the most commonly accepted today.

According to the viewpoint explained above logic is then a prerequisite for mathematics and must be distinguished from the subject called "mathematical logic" which, in its turn, is a chapter of mathematics itself. Mathematical logic models (mirrors) logic within mathematics. This viewpoint was initiated by Hilbert who was one of the main originators and proponents of formalism. Hilbert's program was to prove, within mathematics, that mathematics does not lead to contradictions (it is consistent) and can prove or disprove any of its statements (it is complete). This program received a serious blow from Gödel who proved in particular two mathematical theorems whose translations into English are as follows: 1) consistency of mathematics cannot be proved within mathematics; and 2) if mathematics is consistent then it is incomplete. On the other hand work by many people, including Gödel, showed that, in spite of the failure of Hilbert's original program, the formalist approach can be successfully pursued provided goals less ambitious than Hilbert's are being put forward. Our approach in this course will be along the general lines of Hilbert's formalism although there will be significant deviations here from Hilbert's ideology, as we shall explain below.

Note that mathematical logic has logic as a prerequisite: in order to even talk about mathematical logic one needs mathematics (i.e., set theory) which is in its turn grounded in logic. If we use the symbol $\subset$ to indicate containment then schematically one can represent the above discussion as follows:

$$\{mathematical\ logic\} \subset \{mathematics\} \subset \{logic\}.$$

2. Ontological assumptions. Ontology is about the question of existence (or reality). And, at least in a correspondence theory of truth, the concept of truth is defined as "agreement with reality." All of this becomes problematic for mathematics because the ontological status of mathematical objects seems to be different from that of the physical objects of our immediate experience.

One approach to the ontology of mathematics is to postulate a "mathematical world" that has "states of affairs"; and one can declare that statements about this world are either the case (in which case they are said to be true) or are not the case (in which case they are said to be false). The mathematical world needs to be "infinite" if anything non-trivial is to be achieved. Here "infinity" is taken in a vague, non-mathematical sense; but, in any case, it is viewed as "completed infinity" or "actual infinity" (infinity that is identical to itself throughout time or, rather, is outside time) rather than "potential infinity" (i.e., finiteness in discrete form that, at any moment in time, can grow). Such a world was postulated by the ancient Greek geometers and consisted of infinite space in which an infinity of "perfect" geometric figures "exist" in eternity. This world was an integral part of (and indeed the prototype for) the platonic ontology of the universals. The Greek paradigm ruled unchallenged until, in modern times, the necessity arose for a rethinking of the foundations. A new proposal for a "mathematical world" was put forward by Cantor in the form of his set theory. According to Cantor's original "definition," a set is a "collection into a whole M of definite and separate objects m of our intuition or our thought." The world of Cantor sets is a world of "completed infinities" (indeed there is a whole hierarchy of infinities in this world); this can be seen as a "maximal" ontology that prescribes no limits to existence. Subsequently more restricted versions of set theory were proposed (e.g., by von Neumann); but the semantics of these versions still requires an ontology of infinity.

On the contrary the (formalist inspired) ontology we will assume in this course will be "minimal": the only things admitted into existence will be "finitely many" symbols and their various combinations called texts. (Strictly speaking, along with texts, one would also need to admit into existence a subject, and its ability and will to operate with texts; we will completely ignore the subject in what follows.) Texts are viewed as physical objects of our immediate experience (e.g., written texts). All texts are "finite." Admittedly more and more texts can be produced (so there is a potential infinity implicit in text production); but at every moment there are only "finitely many" texts hence there are definitely no "completed infinities" in the world of texts. This course will be entirely about texts and, in its turn, it is a text. We operate with texts "as if" they were "about the mathematical world" (e.g., "about the world of Cantor sets or von Neumann sets"); but, as we shall see, operating with texts does not require pretending that texts are "about" anything.

We will actually ignore later the "things" that texts are about unless those things are themselves texts. The latter caveat is, however, harmless because the "world of texts" itself is, at any moment, finite. Such an ontology is too minimal to be useful in dealing with more general philosophical questions, including those related to natural sciences; but it turns out to be more than enough for dealing with mathematics.

3. Contents of the course. Part 1 of the course is about the part of logic necessary to introduce mathematics; to stress the independence of this part from any mathematical paradigm we will call this part "Pre-mathematical logic." Part 2 is about mathematics. Part 3 is about mathematical logic. This 3 layered structure (corresponding to the 3 Parts above) is a somewhat non-standard way to approach the subject. Hilbert, for instance, insisted that logic and arithmetic should be developed simultaneously; this is not the case in the present text where a first layer of logic (about which many things can be affirmed but very little can be "metaproved") is introduced in Part 1 while the integers and their arithmetic are introduced in Part 2. Finally our Part 3 is viewed as mere part of mathematics rather than a *metamathematics* in the sense of Hilbert. In our paradigm, for instance, the theorems of Gödel have no reference and no meaning; in particular they say nothing about mathematics (unless we translate them into Metalanguage; see below). This is a rather extreme brand of formalism which is "more minimalist" than the usual brands and therefore less ambitious; for instance we will choose to ignore meaning, reference, and truth of mathematical sentences; we will also disallow metaproofs of consistency or completeness. What is being achieved is, in some sense, an elimination of both psychology and metaphysics from the principles that justify mathematics; physics is also eliminated (with the exception of the languages themselves which are viewed as part of the physical world).

Let us discuss in some detail the main ideas of Part 1. The idea that rational thought obeys definite laws goes back to the classics of logic: Aristotle, Leibniz, and Boole. Now rational thought does not operate with the phenomena but rather with names/symbols given/attached to various phenomena. Symbols are concrete physical objects: they are spoken, written, or shown. They are organized in systems of symbols by certain organizational principles. In particular symbols may be combined to generate *sentences*; and sentences may be combined to generate *theories*, etc. All these combinations are referred to as *texts*. The collection of all symbols in a system of symbols is referred to as a *language*. So in some sense logic is an analysis of languages; as such it can be viewed as part of linguistics:

$$\{logic\} \subset \{linguistics\}.$$

Linguistics is also interested in non-logical aspects of languages such as: morphology, phonology, etymology, psychological and neurological aspects, etc. We will not touch upon these here.

There are various types of correspondences between languages; they attach to fragments of one language fragments of another language. A familiar type of correspondences are translations such as:

$$\text{English} \;\rightarrow\; \text{French}$$

which have an interpretative role. But there are other types of correspondences playing various other (combinations of) roles: referential, descriptive, prescriptive, generative, etc.

The main (logical) aspects of languages are: *syntax, semantics, reference, inference, truth.* Syntax is the set of rules governing the way sentences are assembled from symbols; in other words sentences are the syntactically correct strings of symbols. Syntactic rules are formulated in terms of syntactic categories; examples of syntactic categories are *logical categories* (such as *constants, variables, functional symbols, relational predicates, connectives, quantifiers, etc.*) or *grammatical categories* (such as *nouns, verbs, adjectives, etc.*) Natural languages are analyzable from both the logical and the grammatical angles; but mathematics requires logical rather than grammatical categories. So we will essentially ignore the grammatical categories. Semantics (which is concerned with *meaning*) is about what sentences "say." Meaning is a subtle (and controversial) concept in the philosophy of language but in this book we will adopt a minimal viewpoint on the matter; indeed, for us, the meaning of a sentence in a language will be simply defined as the totality of its available translations into other languages. This minimal approach to meaning will be more than enough for developing mathematics because we will choose later to mostly ignore the meaning of mathematical sentences. Reference (or the universe of discourse) consists of "whatever there is in the physical or imaginary worlds that the symbols refer to." Inference is a process by which we accept "new" declarative sentences based on already accepted "old" declarative sentences; there are other types of sentences such as interrogative or imperative that do not need acceptance hence inference. Also there are sentences which are impossible to infer and whose negation is also impossible to infer from anything that is being accepted. What inference consists of is a matter of convention: it can mean anything from "some evidence to believe" to "formal proof." Truth is a property of declarative sentences that have a meaning; a theory of truth usually requires that any such sentence be either true or false; a theory of truth does not require in principle that truth be capable of being inferred: a sentence can be in principle true "objectively" without the possibility of inference. (Not all theories of truth agree here and, in this course, we will actually very soon give up the concept of truth altogether.) As an illustration of the above concepts consider, for instance, the following utterances:

I. *through passes not electron slit*
II. *colorless green ideas sleep furiously*
III. *it is going to be here*
IV. *some men are immortal*
V. *Napoleon was alive at the battle of Waterloo*
VI. *the universe is infinite.*

In the above I is syntactically incorrect so it is not a sentence. II is a sentence; it is an example of Chomsky's and it has some meaning because it possesses translations into various natural languages; but it has a rather "weak meaning" because it has no reasonable translations into more specialized (scientific) languages (where there is no predicate that would translate "sleeping furiously", for instance); also there is no reasonable theory of truth which would make II either true or false. III is a sentence; it has an imprecise reference because "*it*" is ambiguous; it has a meaning once it is translated into a language that we understand; it is susceptible of being true or false. IV and V are sentences with meaning and reference; IV is false and V is true in any reasonable theory of truth; the falsehood of IV and the truth of V are trivially inferred from generally accepted sentences. VI is a sentence, it has meaning and reference, so it is susceptible of being true or false; its truth or falsehood can be thought of as being independent of our ability to infer it (although not all theories of truth agree on this). The above are just examples; one task of the philosophy of language is to define the concepts of syntax, semantics, reference, inference, truth, and build an explanatory theory of language and understanding based on the corresponding definitions.

Going back to reference (or the universe of discourse), one can distinguish two kinds of reference: *linguistics reference* and *non-linguistic reference*. V above refers to the physical "man Napoleon" (whatever that means) rather than to the "word Napoleon"; such a reference can be called non-linguistic. On the other hand saying that a language $\widehat{L}$ has language L as its linguistic reference means that $\widehat{L}$ "talks about" L as a language, i.e., about the symbols, syntax, semantics, etc. of L (or group of languages like L). In this case, once we fixed $\widehat{L}$ and L we may call L the *object language* and $\widehat{L}$ the *metalanguage*. This designation is relative, i.e., a language may be an object language in one designation and a metalanguage in another designation. But once such a designation has been made we will tend to treat object languages and metalanguages according to different standards, as we shall explain momentarily. Linguistic reference is a correspondence between metalanguage and object language:

$$\text{Metalanguage} \quad \rightarrow \quad \text{Object language.}$$

This kind of correspondence introduces a *hierarchy* among languages which is reminiscent of Russell's theory of types; we will not discuss that theory here. The necessity of introducing the object language / metalanguage hierarchy (and indeed metametalanguage, metametametalanguage, etc.) was recognized by Tarski; but our metalanguage will be slightly different from Tarski's (ours will be poorer than Tarski's, as we shall explain later.)

Sentences in metalanguage will be called *metasentences*; meaning in metalanguage will be called *metameaning*, etc. Assuming we have fixed an object language and a metalanguage we will adopt the following principles (which are appropriate for introducing logic and mathematics although are not all appropriate for introducing other sciences or analyzing natural languages): 1) carefully distinguish between the symbols of object language and the same symbols viewed as belonging to metalanguage; 2) ignore the meaning and reference of sentences in the object language; 3) take into consideration the meaning and reference of metasentences in metalanguage; 4) control syntax of sentences in the object languages (as well as translations between various object languages) through metasentences in metalanguage; 5) adopt a self-referential approach to controlling the syntax of metalanguages; this is in order to avoid the introduction of a metametalanguage, etc.; 6) inference of sentences in the object language will be called *proof* and will be a rigid process whose rules are spelled out in metalanguage; inference of metasentences in metalanguage will be called *metaproof* and will obey less strict rules (which should be spelled out, again, in metalanguage, or if ambiguity arises, in metametalanguage); metaproofs must be finitistic (in the sense of Hilbert and Weyl i.e., "reducible to a quantifier free formulation") and should be reducible to verifying (via case by case inspection) that rules and definitions are applied correctly. In particular no metasentence involving quantifiers "over infinite domains" can be metaproved; 7) ban truth predicates (e.g., the words *true* and *false*) from both the object language and the metalanguage. In case we use these words, which we will sometimes do for convenience, these words will be declared meaningless.

By 1 and 7 above we dispose of a class of classical paradoxes. Indeed consider the "liar's paradox" which is the following sentence in English:

"the sentence that you are reading now is false."

If one assumes the sentence is true it looks like its content says it is false. And if one assumes the sentence is false it looks like its content says it is true. There are at least two sources of the above paradox. One is the use of the words *true* and *false* equipped with their meaning. Another one is the identification of English as object language with English as metalanguage. In this course we will adopt, as already mentioned, a viewpoint which eliminates both sources of the above paradox; cf. 1 and 7 above.

Another reason why we want to operate with an object language and a metalanguage level is that object languages (like English, theoretical physics, mathematics) usually have a complicated structure and their meaning is acquired through translations into other complicated object languages (like history/literature, experimental physics, philosophy); whereas the metalanguage that controls their syntax has a relatively simple structure (it refers to these object languages alone, hence, essentially, to marks on a piece of paper) and its meaning is acquired through translation into other simple languages (like the ones governing simple children games or elementary computer programs). Think of the meaning and possibility of inference of the sentence

1) *"radiation is quantized"*

in English compared to the metasentence in "MetaEnglish":

2) the word *"radiation"* occurs in the sentence *"radiation is quantized."*

The meaning of (and so the possibility to infer) the sentence 1 is problematic: it depends on the translation of the sentence into the language of theoretical physics; there may be several translations into one such language and there are several physical theories to choose from; deciding which theory to use depends on translations between various theoretical physics languages and the language of data coming from experimental physics; the decision may depend also upon translations into usual English and interactions of these with sentences expressing extra-scientific issues, etc. On the other hand the metameaning of (and the possibility to infer) the metasentence 2 above is arguably less problematic: the metameaning depends on the language describing the search of a word in a sentence and as such 2 has an obvious meaning and is trivial to infer. This being the case one can defer the analysis of meaning and inference from a complicated/problematic level to an arguably simple/unproblematic level. By the way, the meaning and inference of 2 is not as unproblematic as it seems: cf. the work of Wittgenstein, for instance.

Here are a few more remarks on *inference* (also called *deduction*). If truth were allowed as a predicate, inference would be required to be such that sentences inferred from true sentences must themselves be true. However, since truth of sentences is not defined, we should view inference as a *substitute for the guarantee of truth*. To explain how inference works let us fix, in our discussion, an object language and a metalanguage. A collection of sentences in the object language obtained via inference from a given list of sentences called *specific axioms* will be called a *theory*. The sentences of a theory will be called *theorems*. Inference is defined in terms of sentences referred to as *background axioms*. The process of inference of sentences will be called *proof*; technically proofs are strings of sentences formed according to certain rules. A key role in the handling of quantifiers within proofs will be played by a certain class of *constants* called *witnesses*; this way of proceeding goes back to Hilbert and Weyl. The word *witness* used here is borrowed from model theory but our use of witnesses is *not* model theoretic (it precedes, as in Hilbert, set theory). Proofs are usually presented not as texts in the object language but rather as translations of such texts into yet another language called *Argot*; the latter is a mixture of the object language with the English (or any other natural) language; this procedure facilitates the reading of proofs (but in some sense reintroduces ambiguities).

Part 2 of the course is devoted to mathematics. Mathematics will be introduced as being the same as a specific theory (in the sense of logic) called set theory. The first move will then be to introduce the axioms of set theory (called the ZFC axioms) and to introduce sets themselves which will be defined as the constants of set theory. (In one version of set theory there are no other sets except the witnesses.) Then we will introduce some of the basic concepts of set theory: maps, relations, operations. The next step will be to introduce the basic types of numbers: integer numbers, rational numbers, real numbers, complex numbers, residue classes modulo p, and the p-adic numbers. (Here p is any prime number.) The collections of such numbers are denoted by $\mathbb{Z}, \mathbb{Q}, \mathbb{R}, \mathbb{C}, \mathbb{F}_p, \mathbb{Z}_p$, respectively. The integers $\mathbb{Z}$ are the most basic numbers; the existence of the integers will follow from the ZFC axioms; then we will introduce the rest of the numbers via standard constructions in set theory. The part of mathematics that abstracts the behavior of numbers is algebra. Geometry then deals with figures. Finally analysis deals with limits, more generally with the infinite (both large and small). The course will provide a quick introduction to some of the basic objects of algebra, geometry, and analysis. These 3 branches of mathematics are closely interconnected; and in each of these branches one is led to consider all the types of numbers referred to above. We will end Part 2 by discussing *mathematical models* (also called here *metamodels*) and *category* theory. Mathematical models are a special type of translations from certain languages into set theory. The translation can be from languages of natural sciences into the language of sets; this is essentially the mechanism by which pure mathematics can become applied mathematics. Sometimes the translation is from set theory into itself; one such case leads to the concept of *universe*. Universes play a key role in category theory; the latter is one of the main unifying concepts in pure mathematics. We will barely scratch the surface of this subject.

Part 3 will be devoted to briefly introducing mathematical logic; this is a somewhat special chapter of mathematics that can be viewed as a chapter of algebra but whose flavor is different from that of mainstream algebra. In mathematical logic one is concerned with mathematical (i.e., set theoretic) concepts that "mimic" the concepts of pre-mathematical logic. In particular (pre-mathematical) semantic concepts will correspond to *model theoretic* concepts (belonging to set theory). (Model theory in mathematical logic is something completely different from the metamodels discussed earlier; but the intuitive idea is similar.) One can define, within set theory, (formal) consistency and completeness; we will end by discussing Gödel's theorems pertaining to these concepts.

As already mentioned our treatment of logic and mathematics in this book is not standard. Indeed books that deal with general logic, linguistics, or philosophy naturally concentrate on what in the present book is treated in Part 1 alone; however such books generally take a different view on semantics, reference, and truth. Their main problem is the relation between language and "reality" while in our case we assume there is no "reality" except language itself. Mathematical books concentrate on what here is covered by Part 2; they generally assume, however, that mathematical sentences are not sentences in set theory (as in the present book) but, rather, have the world of sets as a universe of reference; as such mathematics, in these books, has a well-defined semantics and an absolute notion of truth. Finally books on mathematical logic are mainly interested in what here is covered by Part 3 and hence assume that mathematics is already in place; in such books sets generally play a dual role: they appear as collections of symbols and also as models of theories; in both incarnations sets have the same status as in mathematical books, in particular their world has a semantics and a notion of truth. In particular, sentences such as $a \in A$ are either true or false depending on what a and A "really stand for." Our treatment, on the other hand, will identify mathematics with set theory, and will do away with reference, semantics, and truth in mathematics. For us a sentence such as $a \in A$ will have no truth value but, rather, should be viewed as a sequence of 3 mere symbols. This will conceptually simplify the whole exposition and, in particular, will eliminate any trace of metaphysics from the picture. The price to be payed is the development of the subject in 3 successive steps, as explained above.

4. Suggestions on how to teach this course. If all chapters are thoroughly covered, and if some of the more difficult exercises are discussed in class, there is enough material for a one year course. For a one semester course one should probably proceed as follows. One should start by covering Chapter 1 in some depth. Chapters 2-4 could be covered less thoroughly and their in-depth reading could be assigned as independent study. One could continue by covering Chapters 5-10 in some detail. Chapter 11 is not part of the main narrative but should not be skipped because it illustrates how logic operates in non-mathematical contexts. Chapters 12 and 13 could be taught "in parallel." Chapters 14-22 are an introduction to the basic set theory and number systems and should be covered thoroughly. Chapters 23-42 could be covered in less detail, with some of them assigned as independent study. Chapters 43-48 are, again, directly relevant to the mission of the course and should be taught in some depth. In principle, every chapter (except the first one) depends on previous chapters; so no going back and forth between chapters should be attempted.

Part 1

Pre-mathematical logic

CHAPTER 1

Languages

We are interested in the analysis of language with special emphasis on *syntax, semantics, reference, inference, truth*; this analysis will be referred to as (general) *logic*. We begin with an informal discussion; more details will be given in subsequent chapters of Part 1 of this book. In particular we will discuss later *theories* which are certain collections of sentences in a language. We agree that nothing is available to us except language; in particular we need to analyze language using language. Disentangling this self-referential aspect of the analysis will soon be one of our main concerns. *Mathematics* itself will be defined in Part 2 of the book as a theory (called *set theory*) in a certain language; it will have no semantics, no reference, and no notion of truth. Note that in Part 1 mathematics (set theory) is not available yet so it cannot be used to analyze language; this makes the (general) logic of Part 1 of the book a *pre-mathematical logic*. Once mathematics is available through Part 2 one can revisit the discussion of pre-mathematical logic within mathematics. The resulting discussion is referred to as *mathematical logic* and will be briefly presented in Part 3 of the book. This "extreme formalist" way of organizing the field is not standard; cf the Introduction for a comparison with other approaches.

EXAMPLE 1.1. (Logical analysis) We will introduce here two languages, English and Formal, and we will analyze their interconnections.

Let us start with a discussion of English. The English language is the collection L_{Eng} of all English words (plus separators such as parentheses, commas, etc.). We treat words as individual symbols (and ignore the fact that they are made out of letters). Sometimes we admit as symbols certain groups of words. One can use words to create strings of words such as

    0) *"for all not Socrates man if"*

The above string is considered "syntactically incorrect." The sentences in the English language are the strings of symbols that are "syntactically correct" (in a sense to be made precise later). Here are some examples of sentences in this language:

    1) *"if Socrates is a wise man then Socrates is a man"*
    2) *"Socrates is not a king and Socrates' father is a king"*
    3) *"for all things either the thing is not a man or the thing is mortal"*
    4) *"there exists somebody who is Plato's teacher"*
    5) *"for all things the thing is Socrates if and only if the thing is Plato's teacher"*

We should note right away that "in reality" Socrates' father was *not* a king (but rather a mason); so if we were to define/discuss truth of English sentences then sentence 2 would qualify as false. However the concept of truth has not been addressed yet and we should be prepared to discuss sentences regardless of their apparent "truth value."

In order to separate sentences from a surrounding text we put them between quotation marks (and sometimes we write them in italics). So quotation marks do not belong to the language but rather they lie outside the language. Checking syntax presupposes a partitioning of L_{Eng} into various categories of words; no word should appear in principle in two different categories, but this requirement is often violated in practice (which may lead to different readings of the same text). Here are the categories:

- variables: *"thing, somebody,..."*
- constants: *"Socrates, Plato, the Wise, the Kings,..."*
- functional symbols: *"the father of, the teacher of,..."*
- relational predicates: *"is (belongs to), is a man, is mortal,..."*
- connectives: *"and, or, not, if...then, if and only if"*
- quantifiers: *"for all, there exists"*
- equality: *"is, equals"*
- separators: parentheses *"(,)"* and comma *","*

The above categories are referred to as *logical categories*. (They are quite different from, although related to, the *grammatical categories* of *nouns, verbs*, etc. See Remark 1.15 below for a more in depth discussion of grammatical categories; here we will only allude to them.) In general objects are named by constants or variables. (So constants and variables roughly correspond to nouns.) Constants are names for specific objects while variables are names for non-specific (generic) objects. Relational predicates say/affirm something about one or several objects; if they say/affirm something about one, two, three objects, etc., they are unary, binary, ternary, etc. (So roughly unary relational predicates correspond to intransitive verbs; binary relational predicates correspond to transitive verbs.) Functional symbols have objects as arguments but do not say/affirm anything about them; all they do is refer to (or name, or specify, or point towards) something that could itself be an object. (Functional symbols are sometimes referred to as functional predicates but we will not refer to them as predicates here; this avoids confusion with relational predicates.) Again they can be unary, binary, ternary, etc., depending on the number of arguments. Connectives connect/combine sentences into longer sentences; they can be unary (if they are added to one sentence changing it into another sentence, binary if they combine two sentences into one longer sentence, ternary, etc.). Quantifiers specify quantity and are always followed by variables. Separators separate various parts of the text from various other parts.

In order to analyze a sentence using the logical categories above one first looks for the connectives and one splits the sentence into simpler sentences; alternatively sentences may start with quantifiers followed by variables followed by simpler sentences. In any case, once one identifies simpler sentences, one proceeds by identifying, in each of them, the constants, variables, and functional symbols applied to them (these are the objects that one is talking about), and finally one identifies the functional symbols (which say something about the objects). The above type of analysis (called *logical analysis*) is quite different from the *grammatical analysis* based on the grammatical categories of *nouns, verbs*, etc. (Cf. Remark 1.15 below.)

In our examples of sentences 1-5 logical analysis proceeds as follows.

In 1 *"if...then"* are connectives connecting the simpler sentences *"Socrates is a wise man"* and *"Socrates is a man."* Let us look at the sentence *"Socrates is a wise man."* The word *"Socrates"* here is viewed as a constant; the group of words *"wise man"* is viewed as a constant; *"is"* is a binary relational predicate (it says/affirms something about 2 objects: *"Socrates"* and *"the wise men"*; it says that the first object belongs to the second object). Let us look now at the sentence *"Socrates is a man."* We could read this second sentence the way we read the first one but let us read it as follows: we may still view *"Socrates"* as a constant but we will view *"is a man"* as a unary relational predicate (that says/affirms something about only one object, *Socrates*).

A concise way of understanding the logical analysis of English sentences as above is to create another language L_{For} (let us call it Formal) consisting of the following symbols:

- variables: *"x, y, ..."*
- constants: *"s, p, w, k"*
- functional symbols: *"f, □"*
- relational predicates: *"∈, ρ, †"*
- connectives: *"∧, ∨, ¬, →, ↔"*
- quantifiers: *"∀, ∃"*
- equality: *"="*
- separators: parentheses *"(,)"* and comma *","*

Furthermore let us introduce a rule (called translation) that attaches to each symbol in Formal a symbol in English as follows:

"x, y" are translated as *"thing, somebody"*
"s, p, w, k" are translated as *"Socrates, Plato, the Wise, the Kings"*
"f, □" are translated as *"the father of, the teacher of"*
"∈, ρ, †" are translated as *"belongs to, is a man, is mortal"*
"∧, ∨, ¬, →, ↔" are translated as *"and, or, not, if...then, if and only if"*
"∀, ∃" are translated as *"for all, there exists"*
"=" is translated as *"is"*

Then the English sentence 1 is the translation of the following Formal sentence:

1') *"$(s \in w) \to (\rho(s))$."*

Conversely we say that 1' is a formalization of 1.

Let us continue, in the spirit above, the analysis of the sentences 2,...,5 above.

In 2 *"and"* and *"not"* are connectives (binary and unary respectively): they are used to assemble our sentence 2 from two simpler sentences: *"Socrates is a king"* and *"Socrates' father is a king."* In both these simpler sentences *"Socrates"* and *"king"* are constants; *"the father of"* is a unary functional symbol (it refers to Socrates and points towards/names his father but it does not say anything about Socrates); *"is"* is a binary relational predicate (it says something about two objects). Here is a formalization of 2:

2') *"$(\neg(s \in k)) \wedge (f(s) \in k)$."*

Sentence 3 starts with a quantifier *"for all"* followed by a variable *"things"* followed by a simpler sentence. That simpler sentence is made out of 2 even simpler sentences *"the thing is a man"* and *"the thing is mortal"* assembled via connectives *"not"* and *"or."* Finally *"is a man, is mortal"* are unary relational predicates. Here is a formalization of 3:

3) "$\forall x((\neg\rho(x)) \vee \dagger(x))$."

Sentence 4 starts with a quantifier *"there exists"* followed by a variable *"somebody"* followed by a simpler sentence that needs to be read as *"that somebody is Plato's teacher."* In the latter *"teacher of"* is a unary functional symbol while *"is"* is equality. Here is a formalization of 4:

4') "$\exists y(y = \square(p))$."

Sentence 5 starts with a quantifier *"for all"* followed by a variable *"things"* followed by a simpler sentence. The simpler sentence is assembled from two even simpler sentences: *"the thing is Socrates"* and *"the thing is Plato's teacher"* connected by the connective *"if and only if."* Finally in these latter sentences *"Socrates, Plato"* are constants while *"teacher of"* is a unary functional symbol, and *"is"* is equality. Here is a formalization of 5:

5') "$\forall x((x = s) \leftrightarrow (x = \square(p)))$."

We note that *"or"* in English is disjunctive: *"this or that"* is used in place of *"this or that or both."*

Also note the use of *"is"* in 3 instances: as a binary relational predicate indicating belonging, as part of a unary relational predicate, and as equality.

Note also that we view the variables *"thing"* and *"somebody"* on an equal footing, so we ignore the fact that the first suggests an inanimate entity whereas the second suggests a living entity.

Also note that all verbs in 1-5 are in the present tense. English allows other tenses, of course. But later in mathematics all predicates need to be viewed as tense indifferent: mathematics is atemporal. This is an instance of the fact that natural languages like English have more expressive power than mathematics.

Finally note that the word *"exists"* which could be viewed as a relational predicate is treated instead as part of a quantifier. Sentences like *"philosophers exist"* and *"philosophers are human"* have a totally different logical structure. Indeed *"philosophers exist"* should be read as *"there exists somebody who is a philosopher"* while *"philosophers are human"* should be read as *"for all things if the thing is a philosopher then the thing is a human."* The fact that *"exist"* should not be viewed as a predicate was recognized already by Kant, in particular in his criticism of the "ontological argument."

All of our discussion of English and Formal above is itself expressed in yet another language which needs to be distinguished from English itself and which we shall call Metalanguage. We will discuss Metalanguage in detail in the next chapter (where some languages will be declared *object languages* and others will be declared *metalanguages*). The very latter sentence is written in Metalanguage; and indeed the whole course is written in Metalanguage.

REMARK 1.2. (Naming) It is useful to give names to sentences. For instance if we want to give the name P to the English sentence "*Socrates is a man*" we can write the following sentence in Metalanguage:

$$P \text{ equals } \textit{"Socrates is a man."}$$

So neither P nor the word *equals* nor the quotation marks belong to English; and "*Socrates is a man*" will be viewed in Metalanguage as one single symbol. One can give various different names to the same sentence. In a similar way one can give names to sentences in Formal by writing a sentence in Metalanguage:

$$Q \text{ equals } \textit{"}\rho(s).\textit{"}$$

REMARK 1.3. (Syntax/semantics/reference/inference/truth of languages)

Syntax deals with rules of formation of "correct" sentences. We will examine these rules in detail in a subsequent chapter.

Semantics deals with meaning. For us here the meaning of a sentence will be defined (in a rather minimalist way) as the collection of its available "translations" (where the latter will be understood in a rather general sense). For a person who does not understand English establishing meaning of sentences such as 1,...,5 above requires relating these sentences to sentences in another language (e.g., translating them into French, German, a "picture language," sign language, Braille, etc., or using a correspondence to "deeper" languages, as in the work of Chomsky); the more translations available the more definite the meaning. On the other hand the meaning of 1',...,5' is taken to be given by translating them into the sentences 1,...,5 (where one assumes one knows English).

Reference (or universe of discourse) is "what sentences are about." Words in English may refer to the physical or imaginary worlds (including symbols in languages which are also viewed as physical entities); e.g., the English word "*Socrates*" refers to the physical "*man Socrates*"; the word "*Hamlet*" refers to something in the imaginary world. Metalanguage, on the other hand, refers to other languages such as English or Formal; so the universe of discourse of Metalanguage consists of other languages; such a reference will be called *linguistic reference*. Reference to things other than languages will be called *non-linguistic reference*. Sentences in Formal can be attached a reference once they are translated into English, say; then they have the same reference as their translations.

Inference is a process by which we accept declarative sentences that have a meaning based on other declarative sentences that are already accepted; see the comments below on declarative sentences. There is a whole array of processes that may qualify as inference from belief to mechanical proof.

We could also ask if the sentences 1,...,5, 1',...,5' are "true" or "false." We will not define *truth/falsehood* for sentences in any of our languages. Indeed a theory of truth would complicate our analysis beyond what we are ready to undertake; on the other hand dropping the concepts of truth and falsehood will not affect, as we shall see, our ability to develop mathematics.

REMARK 1.4. (Declarative/imperative/interrogative sentences) All sentences considered so far were declarative (they declare their content). Natural languages have other types of sentences: imperative (giving a command like: "Lift this weight!") and interrogative (asking a question such as: "Is the electron in this portion of space-time?"). In principle, from now on, we will only consider declarative sentences in our languages. An exception to this will later be the use of imperative forms in a language called Argot; Argot will be a language in which we will write most of our proofs and imperative forms to be used in it will be, for instance: "consider," "assume," "let...be," "let us...," etc.)

REMARK 1.5. (Definitions) A language may be enlarged by definitions. More precisely one can add new predicates or constants to a language by, at the same time, recording certain sentences, called definitions. As an example for the introduction of a new relational predicate in English we can add to English the relational predicate *is an astrochicken* by recording the following sentence:

DEFINITION 1.6. Something is an astrochicken if and only if it is a chicken and also a space ship.

Here are alternative ways to give this definition:

DEFINITION 1.7. An astrochicken is something which is a chicken and also a space ship.

DEFINITION 1.8. Something is called (or referred to as) astrochicken if it is a chicken and also a space ship.

Similarly, if in Formal we have a binary relational predicate $\in$ and two constants c and s then one could introduce a new relational predicate ϵ into Formal and record the definition:

DEFINITION 1.9. $\forall x (\epsilon(x) \leftrightarrow ((x \in c) \wedge (x \in s)))$.

The two definitions are related by translating $\in$, c, s, and ϵ as *"belongs to,"* *"chicken,"* *"space ships,"* and *"is an astrochicken,"* respectively. The word *astrochicken* is taken from a lecture by Freeman Dyson.

In a similar way one can introduce new functional symbols or new constants.

In the above discussion we encountered 2 examples of languages that we described in some detail (English and Formal) and one example of language (Metalanguage) that we kept somehow vague. Later we will introduce other languages and make things more precise. We would like to "define" now languages in general; we cannot do it in the sense of Remark 1.5 because definitions in that sense require a language to begin with. All we can do is describe in English what the definition of a language would look like. So the remark below is NOT a definition in the sense of Remark 1.5.

REMARK 1.10. (Description in English of the concept of language) A *first order language* (or simply a *language*) is a collection L of symbols with the following properties. The symbols in L are divided into 8 categories called *logical categories*. They are: variables, constants, functional symbols, relational predicates, logical connectives, quantifiers, equality, and separators. Some of these may be missing. Also we assume that the list of variables and constants may grow indefinitely: we can always add new variables and constants. Finally we assume that the only allowed separators are parentheses $(,)$ and commas; we especially ban quotation marks "..." from the separators allowed in a language (because we want to use them as parts of constants in Metalanguage). Given a language L one can consider the collection L^* of all strings of symbols. We will later define what one means by saying that a string in L^* is a sentence. The collection of sentences in L^* is denoted by L^s. (We sometimes say "sentence in L" instead of sentence in L^*.) As in the examples above we can give names $P, ...$ to the sentences in L; these names $P, ...$ do NOT belong to the original language. A translation of a language L into another language L' is a rule that attaches to any symbol in L a symbol in L'; we assume constants are attached to constants, variables to variables, etc. Due to syntactical correctness (to be discussed later) such a translation attaches to sentences P in L sentences P' in L'. The analysis of translations is part of semantics and will be discussed later in more detail. Actually the above concept of translation should be called *word for word translation* (or *symbol for symbol*) and it is too rigid to be useful. In most examples translations should be allowed to transform sentences into sentences according to rules that are more complicated than the symbol for symbol rule.

REMARK 1.11. (Correspondences between languages) Translations are an example of correspondence between languages. Other examples of correspondences between languages, to be discussed later, are *linguistic reference* and *disquotation*.

EXAMPLE 1.12. (Fixed number of constants) English and Formal are examples of languages. Incidentally in these languages the list of constants ends (there is a "fixed number" of constants). But it is important to not impose that restriction for languages. If instead of English we consider a variant of English in which we have words involving arbitrary many letters (e.g., words like "man," "superman," "supersuperman," etc.) then we have an example of a language with "any number of constants." There is an easy trick allowing one to reduce the case of an arbitrary number of symbols to the case of a fixed number of symbols; one needs to slightly alter the syntax by introducing one more logical category, an *operator* denoted, say, by $'$; then one can form constants $c', c'', c''', ...$ starting from a constant c; one can form variables $x', x'', x''', ...$ from a variable x; and one can do the same with functional symbols, relational predicates, etc.; we will not pursue this in what follows.

EXAMPLE 1.13. (Alternative translations) We already gave an example of translation of Formal into English; cf. Example 1.1. (Strictly speaking that example of translation was not really a word for word translation.) The translation given there for connectives, quantifiers, and equality is called the standard translation. But there are alternative translations as follows.

Alternative translations of $\rightarrow$ into English are *"implies,"* or *"by...it follows that,"* or *"since...we get,"* etc.

An alternative translations of $\leftrightarrow$ into English are *"is equivalent to,"* *"if and only if."*

Alternative translations of $\forall$ into English are *"for any,"* or *"for every."*

Alternative translations of $\exists$ into English are *"for some"* or *"there is an/a."*

English has many other connectives (such as *"before,"* *"after,"* *"but,"* *"in spite of the fact that,"* etc.). Some of these we will ignore; others will be viewed as interchangeable with others; e.g., *"but"* will be viewed as interchangeable with *"and"* (although the resulting meaning is definitely altered). Also English has other quantifiers (such as *"many,"* *"most,"* *"quite a few,"* *"half,"* *"for at least three,"* etc.); we will ignore these other quantifiers.

REMARK 1.14. (Texts) Let us consider the following types of objects:
1) symbols (e.g., $x, y, a, b, f, \Box, ..., \in, \rho, ..., \wedge, \vee, \neg, \rightarrow, \leftrightarrow, \forall, \exists, =, (,))$;
2) collections of symbols (e.g., the collection of symbols above, denoted by L);

$2'$) strings of symbols (e.g., $\exists x \forall y (x \in y)$);

3) collections of strings of symbols (e.g., L^*, L^s encountered above or theories T to be encountered later);

$3'$) strings of strings of symbols (such as the proofs to be encountered later).

In the above, collections are unordered while strings are ordered. The above types of objects $(1, 2, 2', 3, 3')$ will be referred to as *texts*. Texts should be thought of as concrete (physical) objects, like symbols written on a piece of paper or papyrus, words that can be uttered, images in a book or in our minds, etc. We assume we know what we mean by saying that a symbol belongs to (or is in) a given collection/string of symbols; or that a string of symbols belongs to (or is in) a given collection/string of strings of symbols. We will not need to iterate these concepts. We will also assume we know what we mean by performing some simple operations on such objects like: concatenation of strings, deleting symbols from strings, substituting symbols in strings with other symbols, "pairing" strings with other strings, etc. These will be encountered and explained later. Texts will be crucial in introducing our concepts of logic. Note that it might look like we are already assuming some kind of logic when we are dealing with texts; so our introduction to logic might seem circular. But actually the "logic" of texts that we are assuming is much more elementary than the logic we want to introduce later; so what we are doing is not circular.

REMARK 1.15. (Grammatical analysis) Earlier we said that logical analysis of sentences in English is quite different from grammatical analysis. Let us take a quick look at the latter in the following simple example. Consider the following sentence in English:

"the father of Socrates is a king."

The grammatical (as opposed to logical) categories here are:

- nouns: *Socrates, father, king*
- verbs: *is*
- determinators: *a, the*

The sentence (S) above is constructed from a noun phrase (NP) *"the father of Socrates"* followed by a verb phrase (VP) *"is a king."* The noun phrase *"the father of Socrates"* is constructed from the noun phrase *"the father"* and the prepositional phrase (PP) *"of Socrates."* The noun phrase *"the father"* is constructed from a determinator (D) *"the"* and a noun phrase which itself consists of a noun (N) *"father."* The prepositional phrase *"of Socrates"* is constructed from a preposition (P) *"of"* and a noun phrase which itself consists of a noun, *"Socrates."* The verb phrase *"is a king"* is constructed from a verb (V) *"is,"* and a noun phrase *"a king."* The latter is constructed from a determinator (D) *"a,"* and a noun (N) *"king."* One can represent the above grammatical analysis as an array:

S						
NP				VP		
NP		PP		V	NP	
D	N	P	N	V	D	N
the	father	of	Socrates	is	a	king

The above may be referred to as a *grammatical sentence formation*; such a formation is something quite different from the (logical) *formula/sentence formations* to be introduced later based on logical analysis. One can add edges to the array above as follows: each entry X in a given row is linked by an edge to the closest entry Y in the previous row that is above or to the left of X. In this way we get an inverted tree. Such inverted trees are a basic tool in the work of Chomsky on generative grammar, for instance. Alternatively one can encode the information contained in a grammatical sentence formation as follows. One enriches English by adding separators $[_S$ and $]_S$ for sentences, $[_{NP}$ and $]_{NP}$ for noun phrases, etc., and one encodes the above sentence formation as a string of words in this enriched English:

$$[_S[_{NP}[_{NP}[_D\text{the}]_D[_N\text{father}]_N]_{NP}[_{PP}[_P\text{of}]_P[_N\text{Socrates}]_N]_{PP}]_{NP}[_{VP}...]_{VP}]_S.$$

Grammatical sentence formations are obtained by applying substitution rules symbolically written, for instance, as

$S \rightarrow NP\ \ VP$
$NP \rightarrow NP\ \ PP$
$VP \rightarrow V\ \ NP$
$NP \rightarrow D\ \ NP$
$NP \rightarrow N$
$PP \rightarrow P\ \ NP$
$N \rightarrow \text{father}$
$N \rightarrow \text{Socrates}$
etc.

More complicated rules are, of course, present in English. On the other hand very simple "super-rules" that generate these complicated rules in virtually all languages have been discovered. This way of looking at natural languages such as English has deep consequences in psychology and the philosophy of mind; however this approach does not seem appropriate for the study of languages such as Formal or other languages of interest to science and mathematics. Mathematics will require logical (rather than grammatical) analysis. So we will not pursue grammatical analysis beyond this point.

For the next exercises one needs to enrich Formal by new symbols as needed. The translations involved will not be word for word.

EXERCISE 1.16. Find formalizations of the following English sentences:
1) *"I saw a man."*
2) *"There is no hope for those who enter this realm."*
3) *"There is nobody there."*
4) *"There were exactly two people in that garden."*

EXERCISE 1.17. Find formalizations of the following English sentences:
1) *"The movement of celestial bodies is not produced by angels pushing the bodies in the direction of the movement but by angels pushing the bodies in a direction perpendicular to the movement."*
2) *"I think therefore I am."*
3) *"Since existence is a quality and since a perfect being would not be perfect if it lacked one quality it follows that a perfect being must exist."*
4) *"Since some things move and everything that moves is moved by a mover and an infinite regress of movers is impossible it follows that there is an unmoved mover."*

Hints: The word "*but*" should be thought of as "*and*"; "*therefore*" should be thought of as "*implies*" and hence as "*if...then*"; "*since...it follows*" should be thought of, again, as "*implies*."

REMARK 1.18. The sentence 1 above paraphrases a statement in one of Feynman's lectures on gravity. The sentence 2 is, of course Descartes' "cogito ergo sum." The sentence 3 is a version of the "ontological argument" (considered by Anselm, Descartes, Leibniz, Gödel; cf. Aquinas and Kant for criticism). The sentence 4 is a version of the "cosmological argument" (Aquinas).

CHAPTER 2

Metalanguage

In the previous chapter we briefly referred to *linguistic reference* as being a correspondence between two languages in which the first language $\widehat{L}$ "talks about" a second language L as a language (i.e., it "talks about" the syntax, semantics, etc. of L). We also say that $\widehat{L}$ refers to (or has as universe of discourse) the language L. Once we have fixed L and $\widehat{L}$ we shall call L the *object language* and $\widehat{L}$ the *metalanguage*. (The term *metalanguage* was used by Tarski in his theory of truth; but our metalanguage differs from his in certain respects, cf. Remark 2.4 below. Also this kind of correspondence between $\widehat{L}$ and L is reminiscent of Russell's theory of types of which, however, we will say nothing here.)

Metalanguages and object languages are similar structures (they are both languages!) but we shall keep them separate and we shall hold them to different standards, as we shall see below. Sentences in metalanguage are called *metasentences*. If we treat English and Formal as object languages then all our discussion of English and Formal was written in a metalanguage (which is called Metalanguage) and hence consists of metasentences. Let's have a closer look at this concept. First some examples.

EXAMPLE 2.1. Assume we have fixed an object language L such as English or Formal (or several object languages $L, L', ...$). In what follows we introduce a metalanguage $\widehat{L}$. Here are some examples of metasentences in $\widehat{L}$. First some examples of metasentences of the type we already encountered (where the object language L is either English or Formal):

1) x is a variable in the sentence "$\forall x(x \in a)$."
2) P equals *"Socrates is a man."*

Later we will encounter other examples of metasentences such as:

3) $P(b)$ is obtained from $P(x)$ by replacing x with b.
4) Under the translation of L into L' the translation of P is P'.

5) By the table

P	$\neg P$	$P \vee \neg P$
T	F	T
F	T	T

the sentence $P \vee \neg P$ is a tautology.

6) c^P is an existential witness for P.
7) The string of sentences $P, Q, R, ..., U$ is a proof of U.
8) U is a theorem in T.
9) The theory T is consistent.
10) A term is a string of symbols u for which there is a term formation which ends with u.

The metasentences 1, 3, 6 are explanations of syntax in L (see later); 2 is a definition (referred to as a notation or naming); 4 is an explanation of semantics (see later); 5 is part of a metaproof; and 7, 8, 9 are claims about inference (see later). 10 is a definition in metalanguage.

Here are the symbols in $\widehat{L}$:

- variables: symbol, string, language, term, sentence, theorem, $P, Q, R, ...,$ $c_P, c^P, ...$
  - constants: *"Socrates," "Socrates is a man,"* "∧," "=,"..., $T, F, L, L^*, L^s, T, F,$ ..., $P, Q, R, ..., c_P, c^P, ...$
  - functional symbols: the variables in, the translation of, the proof of, $\wedge, \vee, \neg,$ $\rightarrow, \leftrightarrow, \exists x, \forall x,...$
  - relational predicates: is translated as, occurs in, is obtained from...by replacing...with..., is a tautology, is a proof, is consistent,..., follows from, by ... it follows that..., by ... one gets that,...
  - connectives: and, or, not, if...then, if and only if, because,...
  - quantifiers: for all, there exists,...
  - equality: is, equals,...

  - separators: parentheses, comma, period, frames of tables ⊞, ...

REMARK 2.2. Note that names of sentences in the object language become variables (although sometimes also constants) in metalanguage. The sentences of the object language become constants in metalanguage. The connectives of the object language become functional symbols in metalanguage. The symbols "∧, ..." used as constants, the symbols $\wedge, ...$ used as functional symbols, and the symbols *and,...* viewed as connectives should be viewed as different symbols (normally one should use different notation for them).

REMARK 2.3. The above metalanguage can be viewed as a *MetaEnglish* because it is based on English. One can construct a *MetaFormal* metalanguage by replacing the English words with symbols including:
- connectives: & (for and), ⇒ (for if...then), ⇔ (for if and only if)
- equality: ≡ (for is, equals)

We will not proceed this way, i.e., we will always use MetaEnglish as our metalanguage.

REMARK 2.4. What Tarski called metalanguage is close to what we call metalanguage but not quite the same. The difference is that Tarski allows metalanguage to contain the symbols of original object language written *without* quotation marks. So for him (but not for us), if the language is Formal, then the following is a metasentence:

$$\text{"}\forall x \exists y s(x, y)\text{" if and only if } \forall x \exists y s(x, y)$$

Allowing the above to be a metasentence helped Tarski define *truth in a language* (the Tarski T scheme); we will not do this here.

REMARK 2.5. (Syntax/semantics/reference/inference/truth in metalanguage versus object language)

The syntax of object languages will be regulated by metalanguage. On the other hand metalanguage has a syntax of its own which we keep less precise than that of languages so that we avoid the necessity of introducing a metametalanguage which regulates it; that would prompt introducing a metametametalanguage that regulates the metametalanguage, etc. The hope is that metalanguage, kept sufficiently loose from a syntactic viewpoint, can sufficiently well explain its own syntax without leading to contradictions. The very text you are reading now is, in effect, metalanguage explaining its own syntactic problems. The syntactically correct texts in metalanguage are referred to as metasentences. Definitions in metalanguage are called metadefinitions. It is crucial to distinguish between words in sentences and words in metasentences which are outside the quotation marks; even if they look the same they should be regarded as different words.

In terms of semantics sentences in object languages are assumed to have a meaning derived from (or rather identified with) translations into other languages but we will generally ignore this meaning. On the other hand, metasentences have a metameaning derived from their own translations into other languages; we shall assume we understand their metameaning (as it is rather simpler than the meaning of sentences) and we shall definitely *not* ignore it.

Metasentences have a reference: they always refer to sentences in the object language. On the other hand we will ignore the reference of sentences in object language.

Metasentences, as well as sentences in object language, are assumed to have no truth value (it does not make sense to say they are true or false).

For instance the metasentences

a. The word *"elephants"* occurs in the sentence *"elephants are blue."*
b. The word *"crocodiles"* occurs in the sentence *"elephants are gray."*

can be translated in the "metalanguage of letter searches" (describing how to search a word in a sentence, say). Both metasentences have a meaning. Intuitively we are tempted to say that (a) is true and (b) is false. As already mentioned we do not want to introduce the concepts of *true* and *false* in this course. Instead we infer sentences in object language, respectively, metasentences; inference of sentences in object languages will be called *proof*; inference of metasentences in metalanguage will be called *metaproof*. The rules regulating proofs and metaproofs will be explained as we go. As a general rule metaproofs must be finitistic, by which we mean that they are not allowed to involve quantifiers in the metalanguage; in particular we agree that no metasentence involving quantifiers can be metaproved. Metaproving will also be called *checking* or *showing*.

For example we agree that (a) above can be metaproved; also the negation of (b) can be metaproved. Metaproof is usually based on a translation into a "computer language": for instance to metaprove (a) take the words of the sentence *"elephants are blue"* one by one starting from the right (say) and ask if the word is *"elephants"*; the third time you ask the answer is yes, which ends the metaproof of (a). A similar discussion applies to some other types of metasentences; e.g., to the metasentences 1-7 in Example 2.1. The metaproof of 5 in Example 2.1 involves, for instance, "showing tables" whose correctness can be checked by inspection by a machine. (This will be explained later.) The situation with the metasentences 8 and 9 in Example 2.1 is quite different: there is no "finitistic" method (program) that can decide if there is a metaproof for 8; neither is there a "finitistic" method that can find a metaproof for 8 in case there is one; same for 9. But if one already has a proof of U in 8 then checking that the alleged proof is a proof can be done finitistically and this provides a metaproof for 8. Finally 10 is a definition in metalanguage (a metadefinition); definitions will be accepted without metaproofs; in fact most metaproofs consist in checking that a definition applies to a given metasentence. The rules governing the latter would be spelled out in metametalanguage; we will not do this here.

REMARK 2.6. Since P, Q are variables (sometimes constants) in metalanguage and $\wedge, \vee, ..., \exists x$ are functional symbols in metalanguage one can form syntactically correct strings $P \wedge Q, ..., \exists x P$ in metalanguage, etc. If

P equals "$p...$"
Q equals "$q...$"

where $p, ..., q, ...$ are symbols in the object language then:

$P \wedge Q$ equals "$(p...) \wedge (q...)$"

The above should be viewed as one of the rules allowed in metaproofs. Similar obvious rules can be given for $\vee, \exists$, etc. Note that the parentheses are many times necessary; indeed without them the string $P \vee Q \wedge R$ would be ambiguous. We will drop parentheses, however, each time there is no ambiguity. For instance we will never write $(P \vee Q) \wedge R$ as $P \vee Q \wedge R$. Note that according to these conventions $(P \vee Q) \vee R$ and $P \vee (Q \vee R)$ are still considered distinct.

REMARK 2.7. Assume we are given a metadefinition:

P equals "$p....$"

Then we say P is a *name* for the sentence "$p....$" We impose the following rule for this type of metadefinition: if two sentences have the same name they are identical (as strings of symbols in the object language; identity means exactly the same symbols in the same order and it is a physical concept). Note on the other hand that the same sentence in the object language can have different names.

In the same spirit if

$P(x)$ equals "$p...x...$"

is a metadefinition in metalanguage then we will add to the object language a new predicate (still denoted by P) by adding, to the definitions of the object language, the following definition:

$\forall x (P(x) \leftrightarrow (p...x...))$.

So the symbol P appears once as a constant in metalanguage and as a predicate in the object language. (We could have used two different letters instead of just P but it is more suggestive to let P play two roles.) This creates what one can call a correspondence between part of the metalanguage and part of the language. This correspondence, which we refer to as *linguistic reference*, is not like a translation between languages because constants in metalanguage do not correspond to constants in the object language but to sentences (or to new predicates) in the object language. In some sense this linguistic reference is a "vertical" correspondence between languages of "different scope" whereas translation is a "horizontal" correspondence between languages of "equal scope." The words "vertical" and "horizontal" should be taken as metaphors rather than precise terms.

REMARK 2.8. (Disquotation) There is a "vertical" correspondence (called *disquotation* or *deleting quotation marks*) that attaches to certain metasentences in metalanguage a sentence in the object language. Consider for instance the metasentence in MetaEnglish

1) From "*Socrates is a man*" and "*all men are mortal*" it follows that "*Socrates is mortal.*"

Its disquotation is the following sentence in (object) English:

2) "*Since Socrates is a man and all men are mortal it follows that Socrates is mortal.*"

Note that 1 refers to some sentences in English whereas 2 refers to something (somebody) called Socrates. So the references of 1 and 2 are different; and so are their meaning (if we choose to care about the meaning of 2 which we usually don't).

If P equals "*Socrates is a man*," Q equals "*all men are mortal*," and R equals "*Socrates is mortal*" then 2 above is also viewed as the disquotation of:

1') From P and Q it follows that R.

Disquotation is not always performable: if one tries to apply disquotation to the metasentence

1) x is a free variable in "*for all x, x is an elephant*"

one obtains

2) "*x is a free variable in for all x, x is an elephant*"

which is not syntactically correct.

Disquotation is a concept involved in some of the classical theories of truth, e.g., in Tarski's example:

"*Snow is white*" if and only if snow is white.

Since we are not concerned with truth in this course we will not discuss this connection further.

We will often apply disquotation without any warning if there is no danger of confusion.

REMARK 2.9. (Declarative/imperative/interrogative) All metasentences considered so far were declarative (they declare their content). There are other types of metasentences: imperative (giving a command like: "Prove this theorem!," "Replace x by b in P," "Search for x in P," etc.) and interrogative (asking a question such as: "What is the value of the function f at 5?," "Does x occur in P?," etc.). The syntax of metasentences discussed above only applies to declarative metasentences. We will only use imperative/interrogative metasentences in the exercises or some metaproofs (the latter sharing a lot with computer programs); these other types of metasentences require additional syntactic rules which are clear and we will not make explicit here.

EXERCISE 2.10. Consider the following utterances and explain how they can be viewed as metasentences; explain the syntactic structure and semantics of those metasenteces.

1) To be or not to be, that is the question.
2) I do not make hypotheses.

3) The sentence labelled 3 is false.

4) You say yes, I say no, you say stop, but I say go, go, go.

1 is, of course, from Shakespeare. 2 is from Newton. 3 is, of course, a form of the liar's "paradox." 4 is from the Beatles.

From now on we will make the following convention. In any discussion about languages we will assume we have fixed an object language and a metalanguage. The object language will simply be referred to as the "language." So the word "object" will systematically be dropped.

CHAPTER 3

Syntax

We already superficially mentioned syntax. In this chapter we discuss, in some detail, the syntax of languages. (The syntax of metalanguages will be not explicitly addressed but should be viewed as similar.) All the explanations below are, of course, written in metalanguage.

As we saw a language is a collection L of symbols. Given L we considered the collection L^* of all strings of symbols in L. In this chapter we explain the definition of sentences (which will be certain special strings in L^*). Being a sentence will involve, in particular, a certain concept of "syntactic correctness." The kind of syntactic correctness discussed below makes L a *first order language*. There are other types of languages whose syntax is different (e.g., second order languages, in which one is allowed to say, for instance, "for any relational predicate etc...."; or languages whose syntax is based on grammatical categories rather than logical categories; or computer languages, not discussed in this course at all). First order languages are the most natural (and are entirely sufficient) for developing mathematics.

In what follows we let L be a collection of symbols consisting of variables $x, y, ...$, constants, functional symbols, relational predicates, connectives $\wedge$, $\vee$, $\neg$, $\rightarrow$, $\leftrightarrow$ (where $\neg$ is unary and the rest are binary), quantifiers $\forall, \exists$, equality $=$, and, as separators, parentheses $(,)$, and commas. (For simplicity we considered 5 "standard" connectives, 2 "standard" quantifiers, and a "standard" symbol for equality; this is because most examples will be like that. However any number of connectives and quantifiers, and any symbols for them would do. In particular some of these categories of symbols may be missing.) According to our conventions recall that we will also fix a metalanguage $\widehat{L}$ in which we can "talk about" L.

EXAMPLE 3.1. If L has constants $a, b, ...$, a functional symbol f, and a relational predicate $\square$ then here are examples of strings in L^*:

1) $)(x\forall\square\exists aby(\rightarrow$
2) $f(f(a))$
3) $\exists y((a\square x) \rightarrow (a\square y))$
4) $\forall x(\exists y((a\square x) \rightarrow (a\square y)))$

In what follows we will define terms, formulas, and sentences; 1 above will be neither a term, nor a formula, nor a sentence; 2 will be a term; 3 will be a formula but not a sentence; 4 will be a sentence.

The metadefinition below introduces the new unary relational predicate "*is a term formation*" into metalanguage.

METADEFINITION 3.2. A term formation is a string

t

...
s
...
u

of strings in L^* such that for any string s of symbols in the string of strings above (including t and u) one of the following holds:

1) s is a constant or a variable;

2) s is preceded by $s', s'', \ldots$ such that s equals $f(s', s'', \ldots)$ where f is a functional symbol.

METADEFINITION 3.3. A term is a string u for which there is a term formation ending with u.

REMARK 3.4. Functional symbols may be unary $f(t)$, binary $f(t, s)$, ternary $f(t, s, u)$, etc. When we write $f(t, s)$ we simply mean a string of 5 symbols; there is no "substitution" involved here. Substitution will play a role later, though; cf. 3.16.

EXAMPLE 3.5. If $a, b, \ldots$ are constants, $x, y, \ldots$ are variables, f is a unary functional symbol, and g is a a binary functional symbol, all of them in L, then

$$f(g(f(b), g(x, g(x, y))))$$

is a term; a term formation for it is

b
x
y
$f(b)$
$g(x, y)$
$g(x, g(x, y))$
$g(f(b), g(x, g(x, y)))$
$f(g(f(b), g(x, g(x, y))))$

The latter should be simply viewed, again, as a string of symbols.

REMARK 3.6. If $t, s, \ldots$ are terms and f is a functional symbol then $f(t, s, \ldots)$ is a term. The latter is a metasentence; if it is viewed as involving quantifiers applied to variables $t, s, \ldots$ then this metasentence cannot be metaproved. If, however, one views $t, s, \ldots$ as constants in the metalanguage then a metaproof that $f(t, s, \ldots)$ is a term can be done by "showing" as follows. If $t, s, \ldots$ are terms then we have term formations

t'
t''
...
t

and

s'
s''
...
s

Hence we can write a term formation

t'

t''

...

t

s'

s''

...

s

$f(t, s, ...)$

Hence $f(t, s, ...)$ is a term.

METADEFINITION 3.7. A formula formation is a string

P

...

Q

...

R

of strings in L^* such that for any string of symbols Q in the string of strings above (including P and R) one of the following holds:

1) Q equals $t = s$ where t, s are terms.

2) Q equals $\rho(t, s, ...)$ where $t, s, ...$ are terms and ρ is a relational predicate.

3) Q is preceded by Q', Q'' and Q equals one of $Q' \wedge Q''$, $Q' \vee Q''$, $\neg Q'$, $Q' \to Q''$, $Q' \leftrightarrow Q''$.

4) Q equals $\forall x Q'$ or $\exists x Q'$ where Q' precedes Q.

Recall our convention that if we have a different number of symbols (written differently) we make similar metadefinitions for them; in particular some of the symbols may be missing altogether. For instance, if quantifiers are missing from L, then we ignore 4; if equality is missing from L we ignore 1.

METADEFINITION 3.8. A string R in L^* is called a formula if there is a formula formation that ends with R. We denote by L^f the collection of all formulas in L^*.

REMARK 3.9. Relational predicates can be unary $\rho(t)$, binary $\rho(t, s)$, ternary $\rho(t, s, u)$, etc. Again, $\rho(t, s)$ simply means a string of 5 symbols $\rho, (, t, s,)$ and nothing else. Sometimes one uses another syntax for relational predicates: instead of $\rho(t, s)$ one writes $t\rho s$ or $\rho t s$; instead of $\rho(t, s, u)$ one may write $\rho t s u$, etc. All of this is in the language L. On the other hand if some variables $x, y, ...$ appear in a formula P we sometimes write in metalanguage $P(x, y, ...)$ instead of P. In particular if x appears in P (there may be other variables in P as well) we sometimes write $P(x)$ instead of P. Formulas of the form (i.e., which are equal to one of) $\forall x P$, $\forall x P(x)$ are referred to as universal formulas. Formulas of the form $\exists x P$, $\exists x P(x)$ are referred to as existential formulas. Formulas of the form $P \to Q$ are referred to as conditional formulas. Formulas of the form $P \leftrightarrow Q$ are referred to as biconditional formulas.

EXERCISE 3.10. Give a metaproof of the following:

1) if P and Q are formulas then $P \wedge Q$, $P \vee Q$, $\neg P$, $P \to Q$, $P \leftrightarrow Q$ are formulas.

2) if P is a formula then $\forall x P$ and $\exists x P$ are formulas.

(In the above metasentences P, Q are thought of constants in metalanguage; if the above metasentences were viewed as involving quantifiers then they could not be metaproved.)

EXAMPLE 3.11. Assume L contains a constant c, a unary relational predicate ρ, and a unary functional symbol f. Then the following is a formula:
$$(\forall x(f(x) = c)) \to (\rho(f(x)))$$
A formula formation for it is:

$f(x) = c$
$\rho(f(x))$
$\forall x(f(x) = c)$
$(\forall x(f(x) = c)) \to (\rho(f(x)))$

METADEFINITION 3.12. We define the free occurrences of a variable x in a formula by the following conditions:

1) The free occurrences of x in a formula of the form $t = s$ are all the occurrences of x in t together with all the occurrences of x in s;

2) The free occurrences of x in a formula $\rho(t, s, ...)$ are all the occurrences of x in t, together with all the occurrences of x in s, etc.

3) The free occurrences of x in $P \wedge Q$, $P \vee Q$, $P \to Q$, $P \leftrightarrow Q$ are the free occurrences of x in P together with the free occurrences of x in Q. The free occurrences of x in $\neg P$ are the free occurrences of x in P.

4) No occurrence of x in $\forall x P$ or $\exists x P$ is free.

METADEFINITION 3.13. A variable x is free in a formula P if it has at least one free occurrence in P.

EXAMPLE 3.14.
1) x is not free in $\forall y \exists x(\rho(x, y))$.
2) x is free in $(\exists x(\beta(x))) \vee \rho(x, a)$. ($x$ has a "free occurrence" in $\rho(x, a)$; the "occurrence" of x in $\exists x(\beta(x))$ is not "free.")
3) x is not free in $\forall x((\exists x(\beta(x))) \vee \rho(x, a))$
4) The free variables in $(\forall x \exists y(\alpha(x, y, z))) \wedge \forall u(\beta(u, y))$ are z, y.

METADEFINITION 3.15. A string in L^* is called a sentence if it is a formula (i.e., is in L^f) and has no free variables. Note that
1) L^s is contained in L^f;
2) L^f is contained in L^*;
3) all terms are in L^*; no term is in L^f.

METADEFINITION 3.16. If x is a free variable in a formula P one can replace all its free occurrences with a term t to get a formula which can be denoted by $P\frac{t}{x}$. More generally if $x, y, ...$ are variables and $t, s, ...$ are terms, we may replace all free occurrences of these variables by $t, s, ...$ to get a formula $P\frac{ts...}{xy...}$. A more suggestive (but less precise) notation is as follows. We write $P(x)$ instead of P and then we write $P(t)$ instead of $P\frac{t}{x}$. Similarly we write $P(t, s, ...)$ instead of $P\frac{ts...}{xy...}$ We will constantly use this $P(t), P(t, s, ...)$, etc. notation from now on.

Similarly if u is a term containing x and t is another term then one may replace all occurrences of x in u by t to get a term which we may denote by $u\frac{t}{x}$; if we write $u(x)$ instead of u then we can write $u(t)$ instead of $u\frac{t}{x}$. And similarly we may replace two variables x, y in a term u by two terms t, s to get a term $u\frac{ts}{xy}$, etc. We will not make use of this latter type of substitution in what follows.

EXAMPLE 3.17. If P equals "*x is a man*" then x is a free variable in P. If a equals "*Socrates*" then $P(a)$ equals "*Socrates is a man.*"

EXAMPLE 3.18. If P equals "*x is a man and for all x, x is mortal*" then x is a free variable in P. If a equals "*Socrates*" then $P(a)$ equals "*Socrates is a man and for all x, x is mortal.*"

EXERCISE 3.19. Is x a free variable in the following formulas?
1) "$(\forall y \exists x (x^2 = y^3)) \wedge (x$ *is a man*$)$"
2) "$\forall y (x^2 = y^3)$"
Here the upper indexes 2 and 3 are unary functional symbols.

EXERCISE 3.20. Compute $P(t)$ if:
1) $P(x)$ equals "$\exists y (y^2 = x)$" and "t" equals "x^4."
2) $P(x)$ equals "$\exists y (y$ *poisoned* $x)$" and "t" equals "*Plato's teacher.*"

The following metadefinition makes the concept of *definition* in L more precise:

METADEFINITION 3.21. A definition in L is a sentence of one of the following types:
1) "$c = t$" where c is a constant and t is a term without variables.
2) "$\forall x (\epsilon(x) \leftrightarrow E(x))$" where ϵ is a unary relational predicate and E is a formula with one free variable. More generally for several variables, "$\forall x \forall y (\epsilon(x, y) \leftrightarrow E(x, y))$" is a definition, etc.
3) "$\forall x \forall y ((y = f(x)) \leftrightarrow F(x, y))$" where f is a unary functional predicate and F is a formula with 2 free variables; more generally one allows several variables.

If any type of symbols is missing from the language we disallow, of course, the corresponding definitions.

A language together with a collection of definitions is referred to as a language with definitions.

A definition as in 1 should be viewed as either a definition of c (in which case it is also called *notation*) or a definition of t. A definition as in 2 should be viewed as a definition of ϵ. A definition as in 3 should be viewed as a definition of f. More general types of definitions will be allowed later.

REMARK 3.22. Given a language L with definitions and a term t without variables one can add to the language a new constant c and one can add to the definitions the definition $c = t$. We will say that c is (a new constant) defined by $c = t$.

Similarly given a language L with definitions and a formula $E(x)$ in L with one free variable x one can add to the relational predicates of L a new unary relational predicate ϵ and one can add to the definitions the definition $\forall x (\epsilon(x) \leftrightarrow E(x))$. We will say that ϵ is (a new relational predicate) defined by $\forall x (\epsilon(x) \leftrightarrow E(x))$.

One may introduce new functional symbols f by adding a symbol f and definitions of type 3 or of type 1 (such as $c = f(a, b, ...)$, for various constants $a, b, c, ...$).

REMARK 3.23. In Remark 1.15 we discussed grammatical (as opposed to logical) categories/analysis of sentences in natural languages such as English. There is a syntax associated to that point of view as well. (If we call *logical syntax* the syntax discussed before this point in this chapter then what follows in this remark could be referred to as grammatical syntax.) Grammatical syntax proceeds as follows. One introduces grammatical categories such as: nouns (N), verbs (V), adjectives

(A), prepositions (P), determinators (D), etc., and lists of words in each grammatical category. Then one defines a noun phrase (NP) as a noun possibly preceded by adjectives and a determinator and possibly followed by a prepositional phrase (PP); one defines propositional phrases, verb phrases, etc. in a similar way. This scheme can be formalized. What results is a very general theory applicable to virtually all natural languages, not only to English. This kind of grammatical syntax differs from the logical syntax explained before this remark and is not appropriate for introducing mathematics; therefore we will not pursue this grammatical syntax further.

CHAPTER 4

Semantics

We already superficially mentioned semantics. In this chapter we discuss, in some detail, the semantics of (object) languages. Semantics is the analysis of meaning. The concept of meaning can be controversial and it often involves recourse to either psychology or (some brand of) metaphysics. Here we adopt a minimalistic approach: the meaning of a sentence in a language will be defined as the totality of its available translations into other languages. This definition is somewhat unsatisfactory. Indeed we did not say what the meaning of a sentence *is*; we only specified what it *does*, which is to single out a collection of sentences in other languages. However this minimal definition of meaning will be more than enough for our purpose of developing mathematics. This being said we may forget, in what follows, about *meaning* and only care about *translations*.

METADEFINITION 4.1. Let L and L' be two languages. By a translation of L in L' we understand a rule that attaches to any symbol X in L a symbol X' in L'. We say that in our translation X is translated as X' or that X is mapped into X'. We assume the rule "respects" the 8 types of symbols i.e., attaches variables to variables, constants to constants, functional symbols to functional symbols, etc.

REMARK 4.2. By syntactic correctness it is intuitively clear that for any formula P in L if one replaces the symbols X in P by the symbols X' one obtains a formula P' in L'. And the same with sentences in place of formulas. We cannot metaprove this latter metasentence because it involves quantifiers. We say that P is translated (or mapped) into P' or that P' is the translation of P.

One sometimes needs more flexible notions of translation of formulas. Here is such a more flexible notion. (This concept will not play a role until near the end of the course so for now its metadefinition may be skipped.)

METADEFINITION 4.3. Assume $D(x)$ is a formula in L' with one free variable x (which we refer as a "domain"). Assume we are given a translation of L into L'. Replacing all constants and functional symbols X in L by the corresponding symbols X' in L' we get a way to attach to any term t in L a term t' in L'. We define D-translation of formulas in L into formulas in L' by the following rules:

1) The D-translation $(t = s)'$ of $t = s$ is $t' = s'$.
2) The D-translation $(\rho(t, s, ...))'$ of $\rho(t, s, ...)$ is $\rho'(t', s', ...)$.
3) If P', Q' are D-translations of formulas P, Q in L then the D-translation of $P \wedge Q$, $P \vee Q$, $\neg P$, $P \rightarrow Q$, $P \leftrightarrow Q$ are $P' \wedge Q'$, $P' \vee Q'$, $\neg P'$, $P' \rightarrow Q'$, $P' \leftrightarrow Q'$.
4) If $P(x)$ is a formula in L with one free variable x with D-translation $P'(x)$ then the D-translation $(\forall x P(x))'$ of $\forall x P(x)$ is $\forall x(D(x) \rightarrow P'(x))$ and the D-translation $(\exists x P(x))'$ of $\exists x P(x)$ is $\exists x(D(x) \wedge P'(x))$.

This D-translation intuitively asks that the translated sentences "always refer" to things that have "property" (or "domain") D.

REMARK 4.4. There are even more flexible notions of translations. Some of these are obtained by going from one given language "down to a deeper" language via a "vertical" correspondence and then "up to" a third language (or back to the first language) via yet another "vertical" correspondence, such that the resulting "composition" is a "horizontal" correspondence. One such example occurs when one uses natural languages; cf. the work of Chomsky. Another example will occur later when we discuss proofs; we will then use a "vertical" conversion of text from an object language L into a text in a metalanguage $\widehat{L}$ followed by another "vertical" conversion (called *disquotation*) of the text in $\widehat{L}$ back into a different object language L_{argot}; this will produce a "horizontal" correspondence (still referred to as translation) from L into L_{argot}; we don't need to explain this at this point.

REMARK 4.5. Translations should be thought of as concrete objects like dictionaries, tables, or picture books designed to teach foreign languages. Or as tables such as

P	P'
Q	Q'
...	...

REMARK 4.6. If L is any language with connectives $\wedge, \vee, \neg, \rightarrow, \leftrightarrow$, quantifiers $\forall, \exists$, and equality $=$ then the "standard translation" of L into the English language is the one specified in Example 1.1; e.g., $\forall, \exists$ is translated as *"for all, there exists,"* etc. Sometimes, if we translate a language L in English and we want the translation to say that all variables in L refer to something specific in English (say to crocodiles) we use a translation where

$\forall$ is translated as *"for all crocodiles,"*
$\exists$ is translated as *"there exists a crocodile."*

Later we will use *"sets, rings,"* etc. instead of *"crocodiles."*

EXAMPLE 4.7. If L' and L are English and Formal, respectively (cf. Example 1.1) then, as explained in that Example, there is a translation of L into L' such that the sentences $1, ..., 5$ in L are mapped into the sentences $1, ..., 5$ in L', respectively.

EXERCISE 4.8. Assume
P equals "$\forall x((x > 0) \rightarrow (\exists y \exists z \exists u \exists v(x = y^2 + z^2 + u^2 + v^2)))$."
P' equals *"any positive integer is a sum of four squares."*
Find languages L and L' for the sentences P and P' above and find a translation of L into L' such that P is mapped into P'. Hint: We do not need to know at this point what integers are so we need a constant to denote the collection of integers and a predicate that stands for "belongs." By the way this sentence turns out to be a theorem of Lagrange.

EXERCISE 4.9. Assume
P equals "$(K(O)) \wedge (\forall x(K(x) \rightarrow (x = O)))$."
P' equals *"Oswald shot Kennedy and was the only person who shot Kennedy."*
Find languages L and L' for the sentences P and P' above and find a translation of L into L' such that P is sent into P'.

METADEFINITION 4.10. Let L' be the English language and P' be a sentence in English. By a formalization of P' we mean a sentence P in a language L and a translation of L into L' such that P is mapped into P', the variables, constants, and predicates of L are denoted by single marks on paper, and the connectives, quantifiers, and equality of L are $\wedge, \vee, \neg, \rightarrow, \leftrightarrow, \forall, \exists, =$.

EXAMPLE 4.11. In Example 1.1 the sentences 1',...,5' are formalizations of the sentences 1,...,5.

CHAPTER 5

Tautologies

We start now the analysis of inference within a given language (which is also referred to as deduction or proof). In order to introduce the general notion of proof we need to first introduce tautologies; in their turn tautologies are introduced via certain arrays of symbols in metalanguage called tables.

METADEFINITION 5.1. Let T and F be two symbols in metalanguage. We also allow separators in metalanguage that are frames of tables. Using the above plus arbitrary constants (or variables) P and Q in metalanguage we introduce the following strings of symbols in metalanguage (which are actually arrays rather than strings but which can obviously be rearranged in the form of strings). They are referred to as the truth tables of the 5 standard connectives.

P	Q	$P \wedge Q$
T	T	T
T	F	F
F	T	F
F	F	F

P	Q	$P \vee Q$
T	T	T
T	F	T
F	T	T
F	F	F

P	Q	$P \rightarrow Q$
T	T	T
T	F	F
F	T	T
F	F	T

P	Q	$P \leftrightarrow Q$
T	T	T
T	F	F
F	T	F
F	F	T

P	$\neg P$
T	F
F	T

REMARK 5.2. If in the tables above P is the sentence "p...." and Q is the sentence "q...." we allow ourselves, as usual, to identify the symbols $P, Q, P \wedge Q$, etc. with the corresponding sentences "p...," "q...," "$(p...) \wedge (q...)$," etc. Also: the letters T and F evoke "truth" and "falsehood"; but they should be viewed as devoid of any meaning.

Fix in what follows a language L that has the 5 standard connectives $\wedge$, $\vee$, $\neg$, $\rightarrow$, $\leftrightarrow$ (but does not necessarily have quantifiers or equality).

METADEFINITION 5.3. Let $P, Q, ..., R$ be sentences in L. By a Boolean string generated by $P, Q, ..., R$ we mean a string of sentences

P

Q

...

R

U

...

...

W

such that for any sentence V among $U, ..., W$ we have that V is preceded by V', V'' (with V', V'' among $P, ..., W$) and V equals one of the following:

$$V' \wedge V'', V' \vee V'', \neg V', V' \rightarrow V'', V' \leftrightarrow V''.$$

EXAMPLE 5.4. The following is a Boolean string generated by P, Q, R:

P
Q
R
$\neg R$
$Q \vee \neg R$
$P \rightarrow (Q \vee \neg R)$
$P \wedge R$
$(P \wedge R) \leftrightarrow (P \rightarrow (Q \vee \neg R))$

EXAMPLE 5.5. The following is a Boolean string generated by $P \rightarrow (Q \vee \neg R)$ and $P \wedge R$:

$P \rightarrow (Q \vee \neg R)$
$P \wedge R$
$(P \wedge R) \leftrightarrow (P \rightarrow (Q \vee \neg R))$

REMARK 5.6. The same sentence may appear as the last sentence in two different Boolean strings; cf. the last 2 examples.

METADEFINITION 5.7. Assume we are given a Boolean string generated by $P, Q, ..., R$. For simplicity assume it is generated by P, Q, R. (When more or less than 3 generators the metadefinition is similar.) The truth table attached to this Boolean string and to the fixed system of generators P, Q, R is the following string of symbols (or rather plane configuration of symbols thought of as reduced to a string of symbols):

P	Q	R	U	...	W
T	T	T	...	...	...
T	T	F	...	...	...
F	T	T	...	...	...
F	T	F	...	...	...
T	F	T	...	...	...
T	F	F	...	...	...
F	F	T	...	...	...
F	F	F	...	...	...

Note that the 3 columns of the generators consist of all 8 possible combinations of T and F. The dotted columns correspond to the sentences other than the generators and are computed by the following rule. Assume V is not one of the generators P, Q, R and assume that all columns to the left of the column of V were computed; also assume that V is obtained from V' and V'' via some connective $\wedge, \vee, ...$ Then the column of V is obtained from the columns of V' and V'' using the tables of the corresponding connective $\wedge, \vee, ...,$ respectively.

The above rule should be viewed as a syntactic rule for metalanguage; we did not introduce the syntax of metalanguage systematically but the above is one instance when we are quite precise about it.

EXAMPLE 5.8. Consider the following Boolean string generated by P and Q:

P
Q
$\neg P$
$\neg P \wedge Q$

Its truth table is:

P	Q	$\neg P$	$\neg P \wedge Q$
T	T	F	F
T	F	F	F
F	T	T	T
F	F	T	F

Note that the generators P and Q are morally considered "independent" (in the sense that all 4 possible combinations of T and F are being considered for them); this is in spite of the fact that actually P and Q may be equal, for instance, to $a = b$ and $\neg(a = b)$, respectively.

METADEFINITION 5.9. A sentence S is a tautology if one can find a Boolean string generated by some sentences $P, Q, ..., R$ such that
1) The last sentence in the string is S.
2) The truth table attached to the string and the generators $P, Q, ..., R$ has only Ts in the S column.

REMARK 5.10. We do not ask, in the metadefinition above, that for any Boolean string ending in S and for any generators the last column of the truth table have only Ts; we only ask this to hold for one Boolean string and one system of generators.

In all the exercises and examples below, $P, Q, ...$ are specific sentences.

EXAMPLE 5.11. $P \vee \neg P$ is a tautology. To metaprove this consider the Boolean string generated by P,

P
$\neg P$
$P \vee \neg P$

Its truth table is (check!):

P	$\neg P$	$P \vee \neg P$
T	F	T
F	T	T

This ends our metaproof of the metasentence saying that $P \vee \neg P$ is a tautology.

Remark that if we view the same Boolean string

P
$\neg P$
$P \vee \neg P$

as a Boolean string generated by P and $\neg P$ the corresponding truth table is

P	$\neg P$	$P \vee \neg P$
T	T	T
T	F	T
F	T	T
F	F	F

and the last column in the latter table does not consist of Ts only. This does not change the fact that $P \vee \neg P$ is a tautology. Morally, in this latter computation we had to treat P and $\neg P$ as "independent"; this is not a mistake but rather a failed attempt to metaprove that $P \vee \neg P$ is a tautology.

EXAMPLE 5.12. $(P \wedge (P \rightarrow Q)) \rightarrow Q$ is a tautology; it is called *modus ponens*. To metaprove this consider the following Boolean string generated by P, Q, R:

P
Q
$P \rightarrow Q$
$P \wedge (P \rightarrow Q)$
$(P \wedge (P \rightarrow Q)) \rightarrow Q$

Its truth table is:

P	Q	$P \rightarrow Q$	$P \wedge (P \rightarrow Q)$	S
T	T	T	T	T
T	F	F	F	T
F	T	T	F	T
F	F	T	F	T

EXERCISE 5.13. Explain how the table above was computed.

EXERCISE 5.14. Give a metaproof of the fact that each of the sentences below is a tautology:
1) $(P \rightarrow Q) \leftrightarrow (\neg P \vee Q)$.
2) $(P \leftrightarrow Q) \leftrightarrow ((P \rightarrow Q) \wedge (Q \rightarrow P))$.

EXERCISE 5.15. Give a metaproof of the fact that each of the sentences below is a tautology:
1) $(P \wedge Q) \rightarrow P$.
2) $P \rightarrow (P \vee Q)$.
3) $((P \wedge Q) \wedge R) \leftrightarrow (P \wedge (Q \wedge R))$.
4) $(P \wedge Q) \leftrightarrow (Q \wedge P)$.
5) $(P \wedge (Q \vee R)) \leftrightarrow ((P \wedge Q) \vee (P \wedge R))$.
6) $(P \vee (Q \wedge R)) \leftrightarrow ((P \vee Q) \wedge (P \vee R))$.

METADEFINITION 5.16.
1) $Q \rightarrow P$ is called the converse of $P \rightarrow Q$.
2) $\neg Q \rightarrow \neg P$ is called the contrapositive of $P \rightarrow Q$.

EXERCISE 5.17. Give a metaproof of the fact that each of the sentences below is a tautology:
1) $((P \vee Q) \wedge (\neg P)) \rightarrow Q$ (modus ponens, variant).
2) $(P \rightarrow Q) \leftrightarrow (\neg Q \rightarrow \neg P)$ (contrapositive argument).
3) $(\neg(P \wedge Q)) \leftrightarrow (\neg P \vee \neg Q)$ (de Morgan law).
4) $(\neg(P \vee Q)) \leftrightarrow (\neg P \wedge \neg Q)$ (de Morgan law).

5) $((P \to R) \land (Q \to R)) \to ((P \lor Q) \to R)$ (case by case argument).
6) $(\neg(P \to Q)) \leftrightarrow (P \land \neg Q)$ (negation of an implication).
7) $(\neg(P \leftrightarrow Q)) \leftrightarrow ((P \land \neg Q) \lor (Q \land \neg P))$ (negation of an equivalence).

REMARK 5.18. 2) in Exercise 5.17 says that the contrapositive of an implication is equivalent to the original implication.

EXERCISE 5.19. (VERY IMPORTANT) Give an example showing that the sentence
$$(P \to Q) \leftrightarrow (Q \to P)$$
is not a tautology in general. In other words the converse of an implication is not equivalent to the original implication. Here is an example in English suggestive of the above: "*If c a human then c mortal*" is not equivalent to "*If c is mortal then c is a human.*"

METADEFINITION 5.20. A sentence P is a contradiction if and only if $\neg P$ is a tautology.

EXERCISE 5.21. Give a metaproof of the fact that the sentence $P \lor \neg P$ is a tautology and the sentence $P \land \neg P$ is a contradiction.

EXERCISE 5.22. Formalize the following English sentences, negate their formalization, and give the English translation of these formalized negations:
1) If Plato eats this nut then Plato is a bird.
2) Plato is a bird if and only if he eats this nut.

CHAPTER 6

Witnesses

We are almost ready to introduce proofs which are the culmination of our presentation of inference. There is one technical piece in the puzzle that is missing and that is the concept of witness which we explain now. This concept is crucial for our treatment of quantifiers in proofs.

Let L be a language possessing quantifiers $\forall, \exists$ (but not necessarily connectives or equality).

METADEFINITION 6.1. A witness assignment in L is a rule that attaches to any formula P with exactly one free variable, x, two constants, c_P and c^P in L. The constant c_P is called the witness for the sentence $\forall x P(x)$; c^P is called the witness for the sentence $\exists x P(x)$. If a witness assignment was fixed in L we also say L is a language with witnesses.

So if P is specified the symbols c_P, c^P are viewed as constants in the language L; however, if P is unspecified, c_P, c^P can be viewed as variables in the metalanguage.

EXAMPLE 6.2. Symbols such as
$$c^{x=a}, c_{x\square a}, c^{\forall y((y=b)\vee(y\square x))}$$
are constants in the language if the language contains $x, y, =, \square, a, b$. In order to avoid indices we will often introduce definitions such as
$$\begin{aligned} c' &= c_{x\square a}, \\ c'' &= c^{x=a}, \\ c''' &= c^{\forall y((y=b)\vee(y\square x))}, \text{ etc.} \end{aligned}$$

REMARK 6.3. The use of the word *witness* will become clear soon, when we talk about proofs. For now the intuitive content of this concept is best explained via translation. Assume that $\forall, \exists$ are translated in English as *"for all elephants"* and *"there exists an elephant"* and assume that $P(x)$ is translated as *"x is blue."* Then c_P should be translated as *"the least blue elephant"*; c^P should be translated as *"the most blue elephant."* We will see later how this works in proofs. The concept of witness was introduced (under a different name) by Hilbert; the name of *witness* is borrowed here from model theory where it has a related (but different) use.

METADEFINITION 6.4. Given a language L_0 there is a canonical way to construct a language with witnesses, L, starting from L_0. Indeed one first adds to L_0 two new constants c_P and c^P for each formula P in L_0^f with exactly one free variable. (Each c_P is viewed as a new symbol; and the same with c^P.) Call L_1 the enlarged collection of symbols. (Note that in case L_0 is a language with definitions we DO NOT add to the definitions of L_0 any new definitions for the new constants!) Then repeat this construction with L_1 in place of L_0; we get a new collection of symbols L_2, etc. Then we let L be the collection of all symbols in $L_0, L_1, L_2, \ldots$.

Then clearly L has a natural witness assignment. We call L the witness closure of L_0.

EXAMPLE 6.5. Let L_0 be the language with variables $x, y, z, ...$, no constants, no functional symbols, one binary relational predicate $\in$, connectives $\wedge, \vee, \neg, \rightarrow, \leftrightarrow$, quantifiers $\forall, \exists$, equality $=$, and separators $(,), ,$. Let L be the witness closure of L_0. (This will play a role later when we introduce set theory.) Here are examples of constants in L:

$$c^{\exists y(x \in y)}, c_{\exists y(x \in y)}, ...$$

$$c^{x \in c_{\exists y(x \in y)}}, c_{x \in c_{\exists y(x \in y)}}, ...$$

The first row belongs to L_1, the second to L_2, etc.

EXERCISE 6.6. Write examples of constants belonging to L_3 and L_4.

METADEFINITION 6.7. Let $P(x, y)$ be a formula with free variables x, y and let $Q(x)$ equal "$\forall y P(x, y)$." If c' is the witness for $\exists x Q(x)$ and c'' is the witness for $\forall y P(c', y)$ then we say that c', c'' are the witnesses for $\exists x \forall y P(x, y)$. One gives a similar metadefinition for witnesses $c', c'', c''', ...$ for sentences $\exists x \exists y \forall z...(...)$, etc.

EXAMPLE 6.8. Explicitly if $P(x, y)$ is a formula with free variables x, y the witnesses c', c'' for $\exists x \forall y P(x, y)$ are given by:

$$
\begin{aligned}
c' &= c^{\forall y P(x,y)}, \\
c'' &= c_{P(c^{\forall y P(x,y)}, y)}.
\end{aligned}
$$

REMARK 6.9. If we deal with languages with witnesses we will always tacitly assume that all translations are compatible (in the obvious sense) with the witness assignments. Compatibility can be typically achieved as follows: if one is given a translation of a language L_0 into a language L_0' then this translation can be extended uniquely to a translation, compatible with witness assignments, of the witness closure L of L_0 into the witness closure L' of L_0'.

REMARK 6.10. The concept of witness will later give a sweeping answer to the problem of existence of mathematical objects. The answer will be completely syntactical. Indeed we have created (or postulated) in the discussion above a constant c^P for each possible formula $P(x)$ with exactly one free variable x; this constant will be viewed later as the "answer to the existence problem for objects with property P." For instance the various sets of mathematics (such as unions, products, power sets, etc., and later the integers) will be later defined as witnesses for various corresponding sentences. We will not need a universe of discourse for set theory. Sets will simply be constants indexed by formulas; hence they will exist as symbols written on paper. *Ontology will be entirely replaced by syntax.*

CHAPTER 7

Theories

Recall that for a language L equipped with the 5 standard connectives we defined a collection of sentences in L called tautologies; cf. 5.9. In what follows (essentially throughout the course) we assume that L is a language equipped with the 5 standard connectives $\wedge, \vee, \neg, \rightarrow, \leftrightarrow$, quantifiers $\forall, \exists$, and equality $=$ and we assume L has definitions and witnesses. We shall define in what follows two more classes of sentences in L called *quantifier axioms* and *equality axioms*, respectively. Using these we shall define theories and proofs.

METADEFINITION 7.1. A quantifier axiom is a sentence of one of the following forms. In what follows P runs through all the formulas $P(x)$ with a free variable x and t runs through the terms without variables.

$P(c_P) \rightarrow (\forall x P(x))$.
$(\forall x P(x)) \rightarrow P(t)$.
$P(t) \rightarrow (\exists x P(x))$.
$(\exists x P(x)) \rightarrow P(c^P)$.

One can write the above sentences in the form:

$P(c_P) \rightarrow (\forall x P(x)) \rightarrow P(t) \rightarrow (\exists x P(x)) \rightarrow P(c^P)$.

REMARK 7.2. To understand the intuitive content of the quantifier axioms involving witnesses let us look again at an example considered earlier. So assume that $\forall, \exists$ are translated in English as *"for all elephants"* and *"there exists an elephant"* and assume that $P(x)$ is translated as *"x is blue."* Let c_P be translated as *"the least blue elephant"*; and let c^P be translated as *"the most blue elephant."* Then the axiom $P(c_P) \rightarrow (\forall x P(x))$ is translated as *"if the least blue elephant is blue then all elephants are blue."* Also the axiom $\exists x P(x) \rightarrow P(c^P)$ is translated as *"if there exists a blue elephant then the most blue elephant is blue."* These translations are reasonable and could be kept in the back of our minds when dealing with witnesses; but the correct way to think about witnesses is to completely forget about their translation into natural languages.

METADEFINITION 7.3. An equality axiom is a sentence of one of the following forms. In what follows t, s, u run through all (n-tuples of) terms without variables, f runs through all n-ary functional symbols, and $P(x)$ runs through all formulas with free (n-tuples of) variables x.

$t = t$.
$(t = s) \leftrightarrow (s = t)$.
$((t = s) \wedge (s = u)) \rightarrow (t = u)$.
$(t = s) \rightarrow (f(t) = f(s))$.
$(t = s) \rightarrow (P(t) \leftrightarrow P(s))$.

It is convenient to collect some of the metadefinitions above into one metadefinition:

METADEFINITION 7.4. Let L be a language with definitions and witnesses. By a background axiom we understand a sentence which is either a tautology, or a definition, or a quantifier axiom, or an equality axiom.

METADEFINITION 7.5. A collection T of sentences in L is a theory if and only if it satisfies the following conditions:
1) T contains all background axioms;
2') (Closure under conjunction) If P, Q are in T then $P \wedge Q$ is in T;
2") (Closure under modus ponens) If P and $P \rightarrow Q$ are in T then Q is in T.
The sentences in a theory are called theorems (in that theory).

REMARK 7.6. A *proposition* is a theorem which is less "important" in the theory. A *lemma* is a theorem which "helps prove" another theorem. One usually reserves the word *theorem* for the most important sentences that belong to a theory. For a while we will not make any distinction between theorems, propositions, and lemmas. But later, in mathematics, we will start making this distinction.

METADEFINITION 7.7. If we fix a collection of sentences $A, B, \dots$ and we refer to them as specific axioms then by an axiom we will understand either a background axiom or a specific axiom.

METADEFINITION 7.8. Fix a collection of specific axioms $A, B, \dots$. Let U be a sentence. By a proof (or a derivation) of U from $A, B, \dots$ we understand a string of sentences

P

Q

R

$\dots$

$\dots$

U

with the following property. Let S be one of $P, Q, R, \dots$. Then one of the following holds:
1) S is an axiom.
2) S is preceded in the list by sentences S' and S'' such that S equals $S' \wedge S''$.
3) S is preceded in the proof by sentences S' and $S' \rightarrow S$.

EXERCISE 7.9. Fix a collection of specific axioms $A, B, \dots$. Give a metaproof of the fact that a sentence U has a proof from $A, B, \dots$ if and only if there is a string of sentences

P

Q

R

$\dots$

$\dots$

U

with the following property. Let S be one of $P, Q, R, \dots$. Then one of the following holds:

1) S is an axiom.

2) S is preceded in the list by (not necessarily different) sentences S' and S'' such that $(S' \wedge S'') \to S$ is an axiom. (We then say S is inferred from S' and S'' or that S follows from S' and S''.)

Strings as above will still be referred to as proofs.

METADEFINITION 7.10. Fix again specific axioms $A, B, \dots$. The collection T of all sentences U for which there is a proof of U from $A, B, \dots$ is called the theory with (or generated by the) specific axioms $A, B, \dots$. This T is denoted by $T(A, B, \dots)$.

In the next two exercises the metasentences to be proved should be viewed as free of quantifiers (otherwise they cannot be metaproved).

EXERCISE 7.11. Metaprove that $T(A, B, \dots)$ is indeed a theory.

EXERCISE 7.12. Metaprove that any theory is of the form $T(A, B, \dots)$.

REMARK 7.13. If one removes the background axioms from a system of specific axioms the theory does not change. So we may (and will) assume our systems of specific axioms do not contain background axioms.

REMARK 7.14. Two different systems of specific axioms $A, B, \dots$ and $A', B', \dots$ can generate the same theory: i.e., $T(A, B, \dots)$ may coincide with $T(A', B', \dots)$.

EXERCISE 7.15. Let T be a theory in L with specific axioms $A, B, \dots$. Explain why any background axiom and any specific axiom is a theorem.

EXERCISE 7.16. Assume we have a theory $T(A, B, \dots)$ and a sentence U. Metaprove that U is a theorem in $T(A, B, \dots)$ if and only if we have a string of sentences

P

Q

R

...

...

U

with the following property. Let S be one of $P, Q, R, \dots$. Then one of the following holds:

1) S is a theorem in $T(A, B, \dots)$;

2) S is preceded in the list by (not necessarily different) sentences S' and S'' such that $(S' \wedge S'') \to S$ is a theorem in $T(A, B, \dots)$.

Strings as above will still be referred to as proofs. In particular this says that we can use previously proved theorems to prove new theorems.

REMARK 7.17. If the theory has, among its theorems (in particular among its axioms), the sentence $\exists x P(x)$ then note that the sentence $P(c^P)$ is a theorem. We refer to c^P as the witness for the theorem $\exists x P(x)$. More generally if $c', c'', c''', \dots$ are the witnesses for the sentence $\exists x \exists y \exists z \dots P(x, y, z, \dots)$ and the latter sentence is a theorem then $P(c', c'', c''', \dots)$ is a theorem.

EXERCISE 7.18. Metaprove the latter assertion.

REMARK 7.19. (Important) If the language L is the witness closure of a language L_0 then the only theorems in L^s that we are morally interested in are the ones that already belong to L_0^s. So, in some sense, one views L as an artificial language created by introducing artificial witnesses. The role of L is to allow one to define the notion of theory/theorem; but what one is really interested in are the theorems that do not contain the artificially introduced witnesses.

EXERCISE 7.20. Explain why if one erases the last sentence in a proof one obtains a proof of the next to the last sentence in the original proof.

EXERCISE 7.21. Explain why if two of the sentences in a proof have the form $S' \to S''$ and $S'' \to S'''$, respectively, then adding $S' \to S'''$ to the proof yields a proof. Hint: $((S' \to S'') \land (S'' \to S''')) \to (S' \to S''')$ is a tautology.

EXERCISE 7.22. Explain why if the sentences S', S'' appear in a proof then adding $S' \land S''$ to the proof yields a proof.

EXERCISE 7.23. Explain why if S' appears in a proof and $S' \to S$ is a tautology then adding S to the proof yields a proof.

EXERCISE 7.24. Explain why if S' and $S' \to S$ appear in the proof then adding S to the proof yields a proof. (Use modus ponens.)

EXERCISE 7.25. Explain why if $P \to R$ and $Q \to R$ appear in a proof then adding $(P \lor Q) \to R$ to the proof yields a proof. (Use the case by case argument.)

METADEFINITION 7.26. A theory T is complete if for any sentence P either P is a theorem or $\neg P$ is a theorem; i.e., either P is in T or $\neg P$ is in T.

METADEFINITION 7.27. A theory is inconsistent if there is sentence P such that both P and $\neg P$ are theorems. A theory is consistent if it is not inconsistent; i.e., if for any sentence P either P is not a theorem or $\neg P$ is not a theorem.

REMARK 7.28. Strictly speaking, since the metadefinitions above involve quantifiers one cannot metaprove that a given theory is consistent or complete. (Metaproofs are not allowed to involve quantifiers.) Dealing with completeness/consistency requires reformulating the problem within mathematics (i.e., in mathematical logic) where more tools are available that can deal with quantifiers. See the last part of the course.

REMARK 7.29. Let T be a theory in a language L. So T is contained in L^s. Let F be the collection of all sentences P in L^s such that $\neg P$ is in T. Finally let N be the collection of all sentences that are neither in T nor in F. There are 3 possible cases:

1) T is inconsistent (and in particular complete). In this case both T and F coincide with L^s and N contains no sentence.

2) T is consistent and incomplete. In this case any sentence is exactly in one of T, F, N and N contains at least one sentence.

3) T is consistent and complete. In this case any sentence is exactly in one of T, F hence N contains no sentence.

In case 3 note that we may define truth/falsehood of sentences syntactically as follows: a sentence P is *true* if it is in T; and it is *false* if it is in F. This is a metadefinition and implies that a sentence is either true or false and cannot be both true and false.

However one can prove (cf. Gödel) a sentence in set theory whose translation into English says that if set theory is consistent then it is incomplete. Set theory will be introduced later and will be identified with "mathematics as a whole";

so the intuitive conclusion will be that there is no reasonable metadefinition of truth/falsehood in mathematics.

Due to the discussion above it is safer and simpler to NOT define truth/falsehood; this is the way we follow in this course.

EXERCISE 7.30. Let T be a theory and let F be the collection of all sentences P such that $\neg P$ is in T. Show that if P and Q are in T then $P \wedge Q$ is in T; and if at least one of P or Q is in F then $P \wedge Q$ is in F. Also if exactly one of P and Q is in N then $P \wedge Q$ is in N. And if both P and Q are in N then $P \wedge Q$ is either in F or in N. The preceding utterances are all metasentences and can be summarized in a $T/F/N$ "table" as follows:

P	Q	$P \wedge Q$
T	T	T
T	F	F
F	T	F
F	F	F
N	T	N
N	F	N
T	N	N
F	N	N
N	N	F or N

The top of this table is entirely analogous to the T/F truth table of $\wedge$. Show that the T/F truth tables of $\vee, \rightarrow, \leftrightarrow, \neg$ have similar analogous $T/F/N$ tables. Note the following difference of status between the T/F truth tables and the $T/F/N$ tables: the T/F tables are just some meaningless graphical symbols in metalanguage; they helped us define proofs and theories; the $T/F/N$ tables, being notation for metasentences, have a meaning.

REMARK 7.31. Assume neither P nor $\neg P$ is in the theory $T(A, B, ...)$. One may attempt to give a metaproof that both $T(P, A, B, ...)$ and $T(\neg P, A, B, ...)$ are consistent. (As usual we assume here that $A, B, ...$ are constants in the metalanguage and not variables!) The metaproof could go as follows. Assume for instance that $T(P, A, B, ...)$ is inconsistent. Let

Q
R
...
$C \wedge \neg C$

be a derivation of $C \wedge \neg C$ from $P, A, B, ...$ Then one notes that

A
B
...
$P \rightarrow Q$
$P \rightarrow R$
...
$P \rightarrow (C \wedge \neg C)$
$(C \vee \neg C) \rightarrow \neg P$
$C \vee \neg C$

$$\neg P$$

is a derivation of $\neg P$ from $A, B, \ldots$ so $\neg P$ is in $T(A, B, \ldots)$, a contradiction.

Such a metaproof should not be accepted, however. Indeed the above argument is not a finitistic one! Rather it mimics (in metalanguage) a proof by contradiction (of a sentence in object language). Proofs by contradiction are legal for sentences in object language and illegal for metasentences in metalanguage. We will discuss proofs by contradiction later. Nevertheless let us accept the above for one moment and see what can be metaproved from it.

EXERCISE 7.32. Metaprove (using the above) that if a consistent theory T is not complete then there exist at least two consistent theories containing T and not coinciding with T.

METADEFINITION 7.33. A theory T is maximal if it is consistent and if any consistent theory containing T coincides with T.

EXERCISE 7.34. Metaprove (using the above) that a consistent theory is complete if and only if it is maximal.

REMARK 7.35. One can ask if any consistent theory is contained in a complete theory. The latter question is a metasentence that involves quantifiers so it cannot be metaproved. The mirror of this in mathematical logic is a theorem (a consequence of what is called Zorn's lemma).

REMARK 7.36. The moral of the last few exercises is that the finitistic requirement for metaproofs is such that very few metasentences of interest can be actually metaproved.

CHAPTER 8

Proofs

Throughout this chapter we fix a language L as in the previous chapter (i.e., equipped with the standard 5 connectives $\wedge, \vee, \neg, \rightarrow, \leftrightarrow$, quantifiers $\forall, \exists$, equality $=$, definitions, and witnesses) and we also fix a theory T in L with (specific) axioms $A, B, \ldots$. Recall the metadefinition of a proof 7.8. In this chapter we want to discuss this concept in some detail, and to give the first examples of proofs.

In order to make proofs more intelligible one usually uses labels and for each line one indicates in parentheses how the line was derived and one uses the following format:

THEOREM 8.1. U.

Proof.
1. P (by...)
2. Q (by...)
3. R (by...)
...
...
635. U (by...)
□

The above should be viewed as a combination between the metasentence:

U *is a theorem with proof:* $P, Q, R, \ldots, U$

and a metaproof of it (which consists of the explanations *(by....),(by...),...*).

Here is an example of a theorem and a proof. It is a theorem in any theory. It is referred to as the quantified modus ponens. Let $P(x)$ and $Q(x)$ be formulas with one free variable x; for each such pair of formulas we have the following:

THEOREM 8.2. $((\exists x P(x)) \wedge (\forall x (P(x) \rightarrow Q(x)))) \rightarrow (\exists x Q(x))$.

Proof.
1. $(\exists x P(x)) \rightarrow P(c^P)$ (quantifier axiom).
2. $(\forall x (P(x) \rightarrow Q(x))) \rightarrow (P(c^P) \rightarrow Q(c^P))$ (quantifier axiom).
3. $((\exists x P(x)) \wedge (\forall x (P(x) \rightarrow Q(x)))) \rightarrow (P(c^P) \wedge (P(c^P) \rightarrow Q(c^P)))$ (by 1, 2).
4. $((P(c^P) \wedge (P(c^P) \rightarrow Q(c^P)))) \rightarrow Q(c^P)$ (tautology: modus ponens).
5. $Q(c^P) \rightarrow \exists x Q(x)$ (quantifier axiom).
6. $((\exists x P(x)) \wedge (\forall x (P(x) \rightarrow Q(x)))) \rightarrow (\exists x Q(x))$ (by 3, 4, 5).   □

EXERCISE 8.3. Explain each step in the above proof.

EXAMPLE 8.4. Theorem 8.2 should be understood as a collection of theorems: for each choice of language and each choice of $P(x)$ and $Q(x)$ one has a specific theorem (and a specific proof). For instance say that L contains binary relational predicates $\dagger, \square$ and c, b are constants. Then one has the following theorem:

$$((\exists x(x \dagger b)) \wedge (\forall x((x \dagger b) \to (x \square c)))) \to (\exists x(x \square c)).$$

The proof of Theorem 8.2 becomes, in this setting:

1. $(\exists(x \dagger b)) \to (c^{x \dagger b} \dagger b)$.
2. $(\forall x((x \dagger b) \to (x \square c))) \to ((c^{x \dagger b} \dagger b) \to (c^{x \dagger b} \square c))$, etc.

Here is another easy example of a theorem which is, again, a theorem in any theory. Let $P(x)$ and $Q(x)$ be formulas with one free variable x; for each such formula we have the following:

THEOREM 8.5. $(\forall x(P(x) \wedge Q(x))) \to (\forall x P(x))$.

Proof.

1. $(\forall x(P(x) \wedge Q(x))) \to (P(c_P) \wedge Q(c_P))$ (quantifier axiom).
2. $(P(c_P) \wedge Q(c_P)) \to P(c_P)$ (tautology).
3. $P(c_P) \to (\forall x P(x))$ (quantifier axiom).
4. $(\forall x(P(x) \wedge Q(x))) \to (\forall x P(x))$ (by 1, 2, 3).

$\square$

EXERCISE 8.6. Explain each step in the above proof.

The following theorem is, again, a theorem in any theory. Let $R(x, y)$ be a formula with two free variables x, y; for each such formula we have the following:

THEOREM 8.7. $(\exists x \forall y R(x, y)) \to (\forall y \exists x R(x, y))$.

Proof.

1. $(\exists x \forall y R(x, y)) \to (\forall y R(c^{\forall y R(x,y)}, y))$ (quantifier axiom).
2. $(\forall y R(c^{\forall y R(x,y)}, y)) \to R(c^{\forall y R(x,y)}, c_{\exists x R(x,y)})$ (quantifier axiom).
3. $R(c^{\forall y R(x,y)}, c_{\exists x R(x,y)}) \to (\exists x R(x, c_{\exists x R(x,y)}))$ (quantifier axiom).
4. $(\exists x R(x, c_{\exists x R(x,y)})) \to (\forall y \exists x R(x, y))$ (quantifier axiom).
5. $(\exists x \forall y R(x, y)) \to (\forall y \exists x R(x, y))$ (tautology plus 1, 2, 3, 4).

$\square$

REMARK 8.8. The converse of Theorem 8.7 is seldom a theorem (unless our theory is inconsistent). Here is an example in English that suggests this. Translate $R(x, y)$ as "*x is the father of y.*" Then Theorem 8.7 is translated as "*if there is a person who is the father of everybody then everybody has a father.*" But the converse of Theorem 8.7 is translated as "*if everybody has a father then there is a person who is everybody's father.*"

EXERCISE 8.9. Prove the following sentences:

1) $(\forall x \forall y R(x, y)) \leftrightarrow (\forall y \forall x R(x, y))$.
2) $(\exists x \exists y R(x, y)) \leftrightarrow (\exists y \exists x R(x, y))$.
3) $(\forall x P(x)) \to (\exists x P(x))$.

The following theorem is, again, a theorem in any theory. Let $P(x)$ be a formula with one free variable x; for each such formula we have the following:

THEOREM 8.10. $(\neg(\forall x P(x))) \leftrightarrow (\exists x(\neg P(x)))$.

Proof.

1. $(\forall x P(x)) \leftrightarrow P(c_P)$ (quantifier axiom).

2. $(\neg(\forall x P(x))) \leftrightarrow (\neg P(c_P))$ (by 1).
3. $(\exists x(\neg P(x)) \leftrightarrow (\neg P(c^{\neg P}))$ (quantifier axiom).
4. $P(c_P) \rightarrow P(c^{\neg P})$ (quantifier axiom).
5. $(\neg P(c^{\neg P})) \rightarrow (\neg P(c_P))$ (by 4).
6. $(\neg P(c_P)) \rightarrow (\neg P(c^{\neg P}))$ (quantifier axiom).
7. $(\neg P(c_P)) \leftrightarrow (\neg P(c^{\neg P}))$ (by 5, 6).
8. $(\neg(\forall x P(x))) \leftrightarrow (\exists x(\neg P(x)))$ (by 2, 3, 7).

$\square$

EXERCISE 8.11. Explain each of the steps of the proof above.

EXERCISE 8.12. Prove the following sentences:
1) $(\neg(\exists x P(x))) \leftrightarrow (\forall x(\neg P(x)))$.
2) $(\neg(\forall x \exists y P(x, y))) \leftrightarrow (\exists x \forall y(\neg P(x, y)))$.

REMARK 8.13. It should be clear what the rule is to negate sentences that start with more quantifiers: negation changes $\forall$ into $\exists$, changes $\exists$ into $\forall$ and changes P into $\neg P$. For example,

$$\neg(\forall x \exists y \exists z \forall u \exists v P(x, y, z, u, v)) \leftrightarrow (\exists x \forall y \forall z \exists u \forall v(\neg P(x, y, z, u, v))).$$

EXAMPLE 8.14. We have

$$\begin{aligned}
\neg((\forall x \exists y P(x, y)) &\rightarrow (\exists z \forall u \forall v R(z, u, v))) \\
&\leftrightarrow \neg(\neg(\forall x \exists y P(x, y)) \vee (\exists z \forall u \forall v R(z, u, v))), \\
&\leftrightarrow (\forall x \exists y P(x, y)) \wedge \neg(\exists z \forall u \forall v R(z, u, v))), \\
&\leftrightarrow (\forall x \exists y P(x, y)) \wedge (\forall z \exists u \exists v(\neg R(z, u, v))).
\end{aligned}$$

EXERCISE 8.15. Explain all the steps in the above example.

EXERCISE 8.16. Prove the following sentences:
1) $(\forall x(P(x) \wedge Q(x))) \leftrightarrow ((\forall x P(x)) \wedge (\forall x Q(x)))$.
2) $(\exists x(P(x) \vee Q(x))) \leftrightarrow ((\exists x P(x)) \vee (\exists x Q(x)))$.

EXERCISE 8.17. Let f and g be two unary functional symbols. Prove the following sentences. These sentences correspond to what later in set theory will read: "*the composition of two surjections is a surjection*" and "*the composition of two injections is an injection.*"
1) $((\forall z \exists y(f(x) = y)) \wedge (\forall y \exists x(g(x) = y))) \rightarrow (\forall z \exists x(f(g(x)) = z))$.
2) $((\forall x \forall y((f(x) = f(y)) \rightarrow (x = y))) \wedge (\forall x \forall y((g(x) = g(y)) \rightarrow (x = y)))) \rightarrow ((\forall x \forall y((f(g(x)) = f(g(y))) \rightarrow (x = y))$.

EXERCISE 8.18. Prove the following sentence:
$(\neg(\exists x P(x))) \rightarrow (\forall x(P(x) \rightarrow Q(x)))$.
The sentence intuitively says that things that don't exist have all the conceivable properties. For instance, *from the fact that unicorns don't exist it follows that any unicorn is 10 feet long*. Here $P(x)$ stands for *x is a unicorn* and $Q(x)$ stands for *x is 10 feet long*.

In the next theorem P is a sentence and $Q(x)$ is a formula with free variable x.

THEOREM 8.19. $(\forall x(P \rightarrow Q(x))) \rightarrow (P \rightarrow (\forall x Q(x)))$.

Proof.
1. $(\forall x(P \rightarrow Q(x))) \rightarrow (P \rightarrow Q(c_Q))$ (quantifier axiom).
2. $Q(c_Q) \leftrightarrow (\forall x Q(x))$ (quantifier axiom).

3. $(P \to Q(c_Q)) \leftrightarrow (P \to (\forall x Q(x))$ (by 2).
4. $(\forall x(P \to Q(x))) \to (P \to (\forall x Q(x)))$ (by 1, 3).   □

REMARK 8.20. Consider the following sentences:

A) $(\forall x P(x)) \to P(t) \to (\exists x P(x)) \to P(c^P)$.
B) $(\neg(\forall x P(x))) \leftrightarrow (\exists x(\neg P(x)))$.
C) $(\forall x(P \to Q(x))) \to (P \to (\forall x Q(x)))$.

A represents the quantifier axioms minus the one involving c_P. B and C are Theorems 8.10 and 8.19, respectively. Conversely if one defines the witness assignment such that c_P is $c^{\neg P}$ then the quantifier axioms follow from A and B. Note that the sentences A (with the axiom on c^P removed), B, and C are taken as the quantifier axioms in Manin's book (Manin 2009); the moral of the discussion is that our quantifier axioms (in a language in which c_P coincides with $c^{\neg P}$) are equivalent to Manin's quantifier axioms plus the axiom on c^P. Roughly, apart from the role of witnesses, the two systems express "the same" principles; but it is important to point out that the two systems belong to two different worlds: the system here belongs to pre-mathematical logic whereas the one in Manin's book belongs to mathematical logic.

EXERCISE 8.21. Let $P(x)$ be a formula with free variable x and let Q be a sentence. Prove the following sentence:
    $(\forall x(P(x) \to Q)) \leftrightarrow ((\exists x P(x)) \to Q)$.

REMARK 8.22. Let L be a language (with witnesses), T a theory with axioms $A, B, \ldots$. For a sentence U consider the following statements:

1) $\neg U$ is not a theorem.

2) U is a theorem; i.e., there is a proof of U. Here we recall that the correctness of a proof can be checked by a machine.

3) There is an algorithm that guarantees that a given machine will decide (at some unspecified point in time) whether or not there is a proof for U.

4) There is an algorithm that guarantees that a given machine will decide (at some unspecified point in time) whether or not there is a proof for U and will print that proof in case there is one.

5) There is an algorithm that guarantees that a given machine will decide (within a known interval of time) whether there is a proof for U and will print such a proof if there is one.

Assertion 2 implies 1 (under the assumption of consistency). Also 5 implies 4. Also 4 implies 3. Finally 1 and 3 imply 2. The above shows that there is a whole hierarchy of concepts related to proof: morally 1 is the weakest and 5 is the strongest. However 5 is essentially never the case, while 1 is essentially never checkable. Among the concepts above, the universally accepted standard of "proof" is provided by 2 which we have therefore called *proof*.

REMARK 8.23. Given a language L we recall that it does not make sense to say that a sentence in L is true (or false). However one sometimes abusively uses the word truth in relation to sentences of L as follows. If P is a sentence in L then by the metasentence

    (We) prove that P is true

we mean

(We) prove P.

Furthermore the metasentence

(We) prove that P is false

means

(We) prove ¬P.

As a variant, the metasentence

(We) give an example of an x such that P(x) is false

means

(We) prove $\exists x(\neg P(x))$.

A.s.o.

REMARK 8.24. If one is given a translation of L into L' (as usual, compatible with witness assignments) then any proof in L yields, in a natural way, a proof in L'. In particular if U is a theorem in L then the corresponding sentence U' in L' is a theorem in L'.

EXERCISE 8.25. Formalize the following English sentences, negate their formalization, and give the English translation of these formalized negations:

1) If Plato eats at least one nut then Plato is a bird.
2) Plato is a bird if and only if he eats at least one nut.
3) For any planet there is a sun such that the planet revolves around that sun.
4) There exists a sun with no planets revolving around it.

CHAPTER 9

Argot

Proofs written in the object language are hard to read. In order to facilitate comprehension one usually translates proofs into an "easier-to-read" language called Argot. In this chapter we explain how this is being done.

METADEFINITION 9.1. Given a language L and a translation of L into the English language L_{Eng} one may construct a new language L_{argot}, called *argotic* L (or simply *argot*, or *slang*), by replacing some of the symbols in L by their corresponding symbols in L_{Eng}. Then there is a natural translation of L into L_{argot}.

The term *argot* is borrowed from (Manin 2009) although we give it here a slightly different meaning. By the way, when we will write proofs and translate from L to L_{argot} we will use a more flexible version of the notion of translation; this more flexible version of translation might deserve a different name but for simplicity will still be called *translation*. In this more flexible translation, sentences may change more dramatically than in usual translation; new sentences may actually appear; others may disappear. Everything is subject to rules but rather than spelling out these rules we learn argot and the more flexible translation by example. The above type of translation has two steps: the first (which we already performed in the previous chapter) is to convert proofs from a string of sentences in L into a string of metasentences containing the sentences themselves plus an explanation of the inference process; the second step is to convert this string of metasentences back into a string of sentences in argotic L via a process involving *disquotation* (removing quotation marks).

Let us see how this works. As before we fix a theory T in L with specific axioms $A, B, \dots$.

REMARK 9.2. Start with a proof:

P

Q

R

...

...

U

Such a proof can be viewed (via disquotation) as a text in the language L. (Recall that disquotation means replacing $P, Q, \dots$ by the sentences in L they stand for, with quotation marks deleted). Let us view the proof above, however, as a text in metalanguage. As already noted such a proof can be made more intelligible by adding labels $1, 2, 3, \dots$ and explanations in parentheses showing how each sentence follows from other sentences; the proof will be then a text in metalanguage that could look like:

1. P (tautology)
2. Q (by axiom E and 1)
3. R (by 1 and 2)
....
....
66. U (by axiom B and 3)

The above proof can be rewritten as a text in metalanguage:

Proof. From E and since P we get that Q. Since P and since Q it follows that R. [...] By axiom B and since R it follows that U. $\qquad\square$

Here and later ... indicates skipped words while [...] indicates skipped sentences.

We now apply disquotation to the latter text, i.e., replace the symbols $E, P, Q, ...$ by the sentences in L that they stand for without quotation marks; in addition one is allowed to further replace some (or even all) of the symbols appearing in $E, P, Q, ...$ by corresponding words in English; e.g., if

E equals "$\forall x \exists y(x \heartsuit y)$"

then the proof above becomes a text in argotic L which could look like:

Proof. By axiom [number or name of the axiom] for all x there is a y such that $x \heartsuit y$; and since ... and ... it follows that ....

The process described above should be viewed as a (generalized concept of) translation of a proof from L into argotic L.

REMARK 9.3. As explained in Remark 8.23 one sometimes abusively introduces in metasentences the empty words *true, false* in reference to sentences. This can be done in argot as well: one may replace the words "*we get*" by "*we get that ... is true*" or "*we get that ... holds*," etc. Then the translation of the above proof in argot reads:

Proof. From axiom E and since P we get that Q is true. Since P and since Q is true it follows that R holds. [...] By axiom B and since R holds it follows that U is true. $\qquad\square$

We prefer not to do so but one encounters this abuse in mathematical texts quite often.

EXAMPLE 9.4. The statement and proof of Theorem 8.2 can be translated into argot as follows:

THEOREM 9.5. *If there exists x such that $P(x)$ and if we know that, for all x, $P(x)$ implies $Q(x)$ it follows that there exists an x such that $Q(x)$.*

Proof. We know there exists c' such that $P(c')$. Also we know that for all constants c we have $P(c) \to Q(c)$. In particular for our c' we have $P(c') \to Q(c')$. Since $P(c')$ it follows that $Q(c')$. This ends the proof. $\qquad\square$

Note that c' in the above proof plays the role of c^P in the original proof.

EXAMPLE 9.6. The statement and proof of Theorem 8.5 can be translated into argot as follows:

THEOREM 9.7. *Assume that, for all x, $P(x) \wedge Q(x)$. Then for all x, $P(x)$.*

Proof. Let c be a constant. We know $P(c) \wedge Q(c)$. In particular $P(c)$. Since c was arbitrary the conclusion follows.   □

Note that unlike with c^P (whose role was played in the previous example by a "specific" constant c') the role of c_P is played by an "arbitrary" constant c. The words "specific" and "arbitrary" are, of course, vague. Vagueness is inherent in argot.

EXERCISE 9.8. Translate the statement and proof of Theorem 8.10 into argot.

REMARK 9.9. If U is of the form $H \to C$ (in which case H is called hypothesis and C is called conclusion) then the sentence U is translated into argot as *If H then C* or *Assume H; then C.* So

Theorem. $H \to C$.

is translated into argot as either of the following:

Theorem. *If H then C.*
Theorem. *Assume H; then C.*

Similarly

Theorem. $H \leftrightarrow C$

is translated into argot as:

Theorem. *H if and only if C.*

REMARK 9.10. As with theorems and their proofs, definitions too can be expressed in argot. We will do this on a regular basis.

REMARK 9.11. Formally metalanguage and argot are similar: they both involve symbols from the original language plus English words. But metalanguage and argot should not be confused. Metalanguage has a meaning that we care about (coming from the fact that it is "about" the syntax, semantics, and proofs of sentences; most of the text in this course is written in metalanguage so we better recognize its meaning). On the contrary argot has a meaning that we ignore; intuitively argot seems to be "about" the constants and variables of the language but it is, in fact, "about nothing." One way to distinguish formally between metalanguage and argot is to enforce the following rule: symbols in the original language appear BETWEEN quotation marks in metalanguage and WITHOUT quotation marks in argot.

CHAPTER 10

Strategies

We discuss here a series of standard strategies to prove theorems.

As before we fix a theory T in L with specific axioms $A, B, \dots$.

EXAMPLE 10.1. There are two strategies to prove a theorem of the form $H \to C$: "direct proof" and "proof by contradiction." A direct proof (in the form of string of sentences) may look like this:

1. $H \to P$ (by axiom B)
2. Q (by axiom A and 1)
3. $H \to R$ (by 1 and 2)

....

....

835. $H \to C$ (by axiom S and 3)

One usually translates the above proof into argot by applying disquotation to the following text; to simplify our exposition we will still refer to the following text as argot.

Proof. Assume H. By axiom B we get P. By axiom A it follows that Q. From the latter and by P we get R. [...] By axiom S and since R it follows that C. □

A proof of $H \to C$ by contradiction may look like this:

1. $(\neg C \wedge H) \to P$ (by axiom S)
2. Q (by axiom V and 1)
3. $(\neg C \wedge H) \to R$ (by 1 and 2)

....

691. $(\neg C \wedge H) \to (A \wedge \neg A)$ (by 357 and 76)
692. $\neg(\neg C \wedge H)$ (by 691)
693. $H \to C$ (by 692)

One usually translates a proof as above into argot as follows.

Proof. Assume $\neg C$ and H and seek a contradiction. By axiom S we get P. By axiom V and P it follows that Q. [...] By [here one needs to explicitly state 357 and 76] we get $A \wedge \neg A$, a contradiction. This ends the proof. □

EXERCISE 10.2. Explain what was used in steps 692 and 693 above.

REMARK 10.3. So proofs by contradiction start by assuming the hypothesis is "true" and the conclusion is "false." Then one seeks a contradiction (which may be a priori unrelated to the hypothesis or the conclusion).

EXAMPLE 10.4. Direct proofs and proofs by contradiction can be given to sentences U which are not necessarily of the form $H \to C$. A direct proof of U

is just a proof that ends with U. A proof of U by contradiction could proceed as follows:

    1. $\neg U \to P$ (by axiom A)
    2. $P \to Q$ (tautology)
    ....
    99. $\neg U \to D \wedge \neg D$ (by 1 and 98)
    100. $(\neg D \vee D) \to U$ (by 99)
    101. U (by 100 and 3)

In argot:

Proof. Assume $\neg U$ and seek a contradiction. Then by axiom A we get P. Hence Q. [...] Hence $D \wedge \neg D$, a contradiction. This ends the proof. □

EXERCISE 10.5. Explain how 100 above was obtained.

EXAMPLE 10.6. Here is a strategy to prove a theorem of the form $(H' \vee H'') \to C$; this is called a case by case proof. It may run as follows:

    1. $H' \to P$ (by axiom A)
    2. $H'' \to Q$ (by axiom B)
    3. $H' \vee H'' \to P \vee Q$ (by 1 and 2)
    ....
    ....
    76. $P \to C$ (by 55 and 56)
    77. $Q \to C$ (by 64)
    78. $P \vee Q \to C$ (by 76 and 77)
    79. $H' \vee H'' \to C$ (by 3 and 78)

One usually translates a proof as above into argot as follows; note that this translation of the proof completely changes the order of the various steps.

Proof. There are two cases: Case 1 is H'; Case 2 is H''. Assume first that H'. Then by axiom A we get P. [...] By [here one needs to explicitly state 55 and 56] and since P we get C. Now assume that H''. By axiom B it follows that Q. [...] By [here one needs to explicitly state 64] and Q we get C. So in either case we get C. This ends the proof. □

EXERCISE 10.7. Explain how $3, 78, 79$ were obtained.

Direct proofs, proofs by contradiction, and case by case proofs can be combined; we shall see this in examples. Here is a generic example:

EXAMPLE 10.8. A proof of U by contradiction plus case by case may look as follows. (In the proof below A is an arbitrary sentence. Deciding what A to use is a matter of art rather than science!)

    1. $(\neg U \wedge A) \to B$ (an axiom)
    2. $B \to C$ (by 1 and axiom...)
    ...
    33. $S \to (Q \wedge \neg Q)$ (by axiom ... and also by 4,32)
    34. $(\neg U \wedge A) \to (Q \wedge \neg Q)$ (by 1,...,33)
    35. $(\neg Q \vee Q) \to (U \vee \neg A)$ (by 34)
    36. $\neg Q \vee Q$ (tautology)
    37. $U \vee \neg A$ (by 35, 36)

38. $(\neg U \wedge \neg A) \to B'$ (an axiom)
39. $B' \to C'$ (by ... and axiom...)

...

100. $S' \to (Q' \wedge \neg Q')$ (by Axiom ... and ...)
101. $(\neg U \wedge \neg A) \to (Q' \wedge \neg Q')$ (by 38,...,100)
102. $(\neg Q' \vee Q') \to (U \vee A)$ (by 101)
103. $\neg Q' \vee Q'$ (tautology)
104. $U \vee A$ (by 102, 103)
105. $(U \vee \neg A) \wedge (U \vee A)$ (by 37, 104)
106. U (by 105)

One usually translates a proof as above into argot as follows.

Proof. Assume $\neg U$ and seek a contradiction. There are two cases: either A or $\neg A$. Assume first A. Then, since $\neg U$ and A, by..., we get B. Hence by ... we get C. [...] Hence by ... we get S. Hence $Q \wedge (\neg Q)$, a contradiction. Now assume $\neg A$. Then, since $\neg U$ and $\neg A$ we get B'. Hence C'. [...] Hence S'. Hence $Q' \wedge (\neg Q')$ which is again a contradiction. This ends the proof.

$\square$

EXERCISE 10.9. Explain each step in the above formal proof.

EXAMPLE 10.10. In order to prove a theorem U of the form $P \leftrightarrow Q$ one may proceed as follows:

1. $P \to S$ (by...)
2. $S \to Q$ (by...)
3. $P \to Q$ (by 1 and 2)
4. $Q \to T$ (by...)
5. $T \to P$ (by...)
6. $Q \to P$ (by 4 and 5)
7. $P \leftrightarrow Q$ (by 3 and 6)

Alternatively, in argot,

Proof. We need to prove two things: first that $P \to Q$; and next that $Q \to P$. To prove $P \to Q$ assume P. By ... we get S. By ... we get Q. So $P \to Q$ is proved. To prove $Q \to P$ assume Q. By ... we get T. By ... we get P. So $Q \to P$ is proved. This ends the proof of the theorem.

$\square$

EXAMPLE 10.11. Sometimes a theorem U has the statement:

The following conditions are equivalent:
1) P;
2) Q;
3) R.

What is being meant is that U is

$$(P \leftrightarrow Q) \wedge (P \leftrightarrow R) \wedge (Q \leftrightarrow R)$$

One proceeds "in a circle" as follows. (We just give the argot version of the proof.)

Proof. It is enough to prove 3 things: first that $P \to Q$; then that $Q \to R$; then that $R \to P$. To prove that $P \to Q$ assume P. By P we get ..., hence we get ..., hence we get Q. To prove that $Q \to R$ assume Q. By Q we get ..., hence we get

..., hence we get R. To prove that $R \to P$ assume R. By R we get ..., hence we get ..., hence we get P.   □

EXERCISE 10.12. Explain why, in the last Remark, it is enough to prove $P \to Q$, $Q \to R$, and $R \to P$ in order to conclude U.

EXAMPLE 10.13. In order to prove a theorem of the form $P \wedge Q$ one usually proceeds as follows:

Proof. We first prove P. By ... and ... it follows that ...; by ... and ... it follows that P. Next we prove Q. By ... and ... it follows that ...; by ... and ... it follows that Q. This end the proof.   □

EXAMPLE 10.14. In order to prove a theorem of the form $P \vee Q$ one may proceed by contradiction as follows:

Proof. Assume $\neg P$ and $\neg Q$ and seek a contradiction. By ... and ... it follows that ...; by ... and ... it follows that $A \wedge \neg A$, a contradiction. This end the proof.   □

EXAMPLE 10.15. In order to prove a theorem of the form $\forall x P(x)$ one may proceed as follows:

Proof. Let c be arbitrary. By ... it follows that ... and hence $P(c)$.   □

EXAMPLE 10.16. In order to prove a theorem of the form $\exists x P(x)$ one may proceed as follows; this is called a proof by example and it applies only to existential sentences $\exists x P(x)$ (NOT to universal sentences like $\forall x P(x)$).

Proof. By ... we know that there exists c such that $Q(c)$. By ... and ... it follows that .... By ... and ... it follows that $P(c)$.   □

In the above c is NOT ARBITRARY but rather an EXAMPLE.

We end by discussing fallacies. A fallacy is a logical mistake. Here are some typical fallacies:

EXAMPLE 10.17. *Confusing an implication with its converse.* Say we want to prove that $H \to C$. A typical mistaken proof would be: Assume C; then by ... we get that ... hence H. The error consists of having proved $Q \to P$ rather than $P \to Q$.

EXAMPLE 10.18. *Proving a universal sentence by example.* Say we want to prove $\forall x P(x)$. A typical mistaken proof would be: By ... there exists c such that ... hence ... hence $P(c)$. The error consists in having proved $\exists x P(x)$ rather than $\forall x P(x)$.

EXAMPLE 10.19. *Defining a constant twice.* Say we want to prove $\neg(\exists x P(x))$ by contradiction. A mistaken proof would be: Assume there exists c such $P(c)$. Since we know that $\exists x Q(x)$ let c be (or define c) such that $Q(c)$. By $P(c)$ and $Q(c)$ we get ... hence ..., a contradiction. The error consists in defining c twice in two unrelated ways: first c plays the role of the specific constant c^P; then c plays the role of c^Q. But c^P and c^Q are not the same. We will see examples of this later; see Exercise 17.28.

EXERCISE 10.20. Give examples of wrong proofs of each of the above types. If you can't solve this now, wait until we get to discuss the integers.

REMARK 10.21. Later, when we discuss induction we will discuss another typical fallacy; cf. Example 18.7.

CHAPTER 11

Examples

We analyze in what follows a few toy examples of theories and proofs of theorems in these theories. Later we will present the main example of theory in this course which is set theory (identified with mathematics itself). As usual the witness assignment in our languages will not be given explicitly.

EXAMPLE 11.1. The first example is what later in mathematics will be referred to as the uniqueness of neutral elements. The language L of the theory has constants $e, f, ...$, variables $x, y, ...$, and a binary functional symbol $\star$. We introduce the following definition in L:

DEFINITION 11.2. e is called a neutral element if
$$\forall x((e \star x = x) \wedge (x \star e = x)).$$

So we added *"is a neutral element"* as a new relational predicate. We do not consider any specific axiom. We prove the following:

THEOREM 11.3. *If e and f are neutral elements then $e = f$.*

The sentence that needs to be proved is of the form: "If H then C." Recall that in general for such a sentence H is called hypothesis and C is called conclusion. Here is a direct proof:

Proof. Assume e and f are neutral elements. Since e is a neutral element it follows that $\forall x(e \star x = x)$. By the latter $e \star f = f$. Since f is a neutral element we get $\forall x(x \star f = x)$. So $e \star f = e$. Hence we get $e = e \star f$. Hence we get $e = f$. □

REMARK 11.4. Note that we skipped some of the "explanations" showing what sentences have been used at various stages of the proof; this is common practice.

EXERCISE 11.5. Convert the above argot proof into the original (non-argotic) language. Here by "convert" we understand finding a proof whose translation in argot is the given argot proof.

Here is a proof by contradiction of the same theorem:

Proof. Assume $e \neq f$ and seek a contradiction. So either $e \neq e \star f$ or $e \star f \neq f$. Since $\forall x(e \star x = x)$ it follows that $e \star f = f$. Since $\forall x(x \star f = x)$ we get $e \star f = e$. So we get $e = f$, a contradiction. □

EXERCISE 11.6. Convert the above argot proof into the original (non-argotic) language.

EXAMPLE 11.7. The next example is related to the "Pascal wager." The structure of Pascal's argument is as follows. If God exists and I believe it exists then I will be saved. If God exists and I do not believe it exists then I will not be saved.

If God does not exist but I believe it exists I will not be saved. Finally if God does not exist and I do not believe it exists then I will not be saved. Pascal's conclusion is that if he believes that God exists then there is a one chance in two that he be saved whereas if he does not believe that God exists then there is a zero chance that he be saved. So he should believe that God exists. The next example is a variation of Pascal's wager showing that if one requires "sincere" belief rather than just belief based on logic then Pascal will not be saved. Indeed assume the specific axioms:

A1) If God exists and a person does not believe sincerely in its existence then that person will not be saved.

A2) If God does not exist then nobody will be saved.

A3) If a person believes that God exists and his/her belief is motivated only by Pascal's wager then that person does not believe sincerely.

We want to prove the following

THEOREM 11.8. *If Pascal believes that God exists but his belief is motivated by his own wager only then Pascal will not be saved.*

All of the above is formulated in the English language L'. We consider a simpler language L and a translation of L into L'.

The new language L contains among its constant p (for Pascal) and contains 4 unary relational predicates g, w, s, r whose translation in English is as follows:

g is translated as "*is God*"

w is translated as "*believes motivated only by Pascal's wager*"

s is translated as "*believes sincerely*"

r is translated as "*is saved*"

The specific axioms are

A1) $\forall y((\exists x g(x)) \wedge (\neg s(y)) \rightarrow \neg r(y))$.

A2) $\forall y((\neg(\exists x g(x))) \rightarrow (\neg r(y)))$.

A3) $\forall y(w(y) \rightarrow (\neg(s(y))))$.

In this language Theorem 11.8 is the translation of the following:

THEOREM 11.9. *If $w(p)$ then $\neg r(p)$.*

So to prove Theorem 11.8 in L' it is enough to prove Theorem 11.9 in L. We will do this by using a combination of direct proof and case by case proof.

Proof of Theorem 11.9. Assume $w(p)$. There are two cases: the first case is $\exists x g(x)$; the second case is $\neg(\exists x g(x))$. Assume first that $\exists x g(x)$. Since $w(p)$, by axiom A3 it follows that $\neg s(p)$. By axiom A1 $(\exists x g(x)) \wedge (\neg s(p)) \rightarrow \neg r(p)$. Hence $\neg r(p)$. Assume now $\neg(\exists x g(x))$. By axiom A2 we then get again $\neg r(p)$. So in either case we get $\neg r(p)$ which ends the proof.

$\square$

EXERCISE 11.10. Convert the above argot proof into the original (non-argotic) language.

EXAMPLE 11.11. The next example is again a toy example and comes from physics. In order to present this example we do not need to introduce any physical concepts. But it would help to keep in mind the two slit experiment in quantum mechanics (for which we refer to Feynman's Physics course, say). Now there are two types of physical theories that can be referred to as *phenomenological* and *explanatory*. They are intertwined but very different in nature. Phenomenological theories are simply descriptions of phenomena/effects of (either actual or possible) experiments; examples of such theories are those of Ptolemy, Copernicus, or that of pre-quantum experimental physics of radiation. Explanatory theories are systems postulating transcendent causes that act from behind phenomena; examples of such theories are those of Newton, Einstein, or quantum theory. The theory below is a baby example of the phenomenological (pre-quantum) theory of radiation; our discussion is therefore not a discussion of quantum mechanics but rather it suggests the necessity of introducing quantum mechanics. The language L' and definitions are those of experimental/phenomenological (rather than theoretical/explanatory) physics. We will not make them explicit. Later we will move to a simplified language L and will not care about definitions.

Consider the following specific axioms (which are the translation in English of the phenomenological predictions of classical particle mechanics and classical wave mechanics, respectively):

A1) If radiation in the 2 slit experiment consists of a beam of particles then the impact pattern on the photographic plate consists of a series of successive flashes and the pattern has 2 local maxima.

A2) If radiation in the 2 slit experiment is a wave then the impact pattern on the photographic plate is not a series of successive flashes and the pattern has more than 2 local maxima.

We want to prove the following

THEOREM 11.12. *If in the 2 slit experiment the impact consists of a series of successive flashes and the impact pattern has more than 2 local maxima then in this experiment radiation is neither a beam of particles nor a wave.*

The sentence reflects one of the elementary puzzles that quantum phenomena exhibit: radiation is neither particles nor waves but something else! And that something else requires a new theory which is quantum mechanics. (A common fallacy would be to conclude that radiation is both particles and waves !!!) Rather than analyzing the language L' of physics in which our axioms and sentence are stated (and the semantics that goes with it) let us introduce a simplified language L as follows.

We consider the language L with constants $a, b, ...$, variables $x, y, ...$, and unary relational predicates p, w, f, m. Then there is a translation of L into L' such that:
- p is translated as "is a beam of particles"
- w is translated as "is a wave"
- f is translated as "produces a series of successive flashes"
- m is translated as "produces a pattern with 2 local maxima"

Then we consider the specific axioms
A1) $\forall x(p(x) \rightarrow (f(x) \wedge m(x)))$.
A2) $\forall x(w(x) \rightarrow (\neg f(x)) \wedge \neg m(x)))$.
Here we tacitly assume that the number of maxima cannot be 1. Theorem 11.12 above is the translation of the following theorem in L:

THEOREM 11.13. $\forall x((f(x) \wedge (\neg m(x))) \rightarrow ((\neg p(x)) \wedge (\neg w(x))))$.

So it is enough to prove Theorem 11.13. The proof below is, as we shall see, a combination of proof by contradiction and case by case.

Proof. We proceed by contradiction. So assume there exists a such that $f(a) \wedge (\neg m(a))$ and $\neg(\neg p(a) \wedge (\neg w(a)))$ and seek a contradiction. Since $\neg(\neg p(a) \wedge (\neg w(a)))$ we get $p(a) \vee w(a)$. There are two cases. The first case is $p(a)$; the second case is $w(a)$. We will get a contradiction in each of these cases separately. Assume first $p(a)$. Then by axiom A1 we get $f(a) \wedge m(a)$, hence $m(a)$. But we assumed $f(a) \wedge (\neg m(a))$, hence $\neg m(a)$, so we get a contradiction. Assume now $w(a)$. By axiom A2 we get $(\neg f(a)) \wedge (\neg m(a))$ hence $\neg f(a)$. But we assumed $f(a) \wedge (\neg m(a))$, hence $f(a)$, so we get again a contradiction. $\square$.

EXERCISE 11.14. Convert the above argot proof into the original (non-argotic) language.

EXERCISE 11.15. Consider the specific axioms A1 and A2 above and also the specific axioms:

A3) $\exists x(f(x) \wedge (\neg m(x)))$.
A4) $\forall x(p(x) \vee w(x))$.

Metaprove that the theory with specific axioms A1, A2, A3, A4 is inconsistent. A3 is translated as saying that in some experiments one sees a series of successive flashes and, at the same time, one has more than 2 maxima. Axiom A4 is translated as saying that any type of radiation is either particles or waves. The inconsistency of the theory says that classical (particle and wave) mechanics is not consistent with experiment. (So a new mechanics, quantum mechanics, is needed.) Note that none of the above discussion has anything to do with any concrete proposal for a quantum mechanical theory; all that the above suggests is the necessity of such a theory.

EXAMPLE 11.16. The next example is a logical puzzle from the Mahabharata. King Yudhishthira loses his kingdom to Sakuni at a game of dice; then he stakes himself and he loses himself; then he stakes his wife Draupadi and loses her too. She objects by saying that her husband could not have staked her because he did not own her anymore after losing himself. Here is a possible formalization of her argument.

We use a language with constants $i, d, ...$, variables $x, y, z, ...$, the relational binary predicate "*owns*," quantifiers, and equality $=$. We define a predicate $\neq$ by $(x \neq y) \leftrightarrow (\neg(x = y))$. Consider the following specific axioms:

A1) For all x, y, z if x owns y and y owns z then x owns z.
A2) For all y there exists x such that x owns y.
A3) For all x, y, z if y owns x and z owns x then $y = z$.

We will prove the following

THEOREM 11.17. *If i does not own himself then i does not own d.*

Proof. We proceed by contradiction. So we assume i does not own i and i owns d and seek a contradiction. There are two cases: first case is d owns i; the second case is d does not own i. We prove that in each case we get a contradiction. Assume first that d owns i; since i owns d, by axiom A1, i owns i, a contradiction. Assume now d does not own i. By axiom A2 we know that there exists j such that j owns i. Since i does not own i it follows that $j \neq i$. Since j owns i and i owns d, by axiom A1, j owns d. But i also owns d. By axiom A3, $i = j$, a contradiction. $\square$

EXERCISE 11.18. Convert the above argot proof into the original (non-argotic) language.

EXAMPLE 11.19. This example illustrates the logical structure of the Newtonian theory of gravitation that unified Galileo's phenomenological theory of falling bodies (the physics on Earth) with Kepler's phenomenological theory of planetary motion (the physics of "Heaven"); Newton's theory counts as an explanatory theory because its axioms go beyond the "facts" of experiment. The language L in which we are going to work has variables $x, y, ...$, constants S, E, M (translated in English as "Sun, Earth, Moon"), a constant R (translated as "the radius of the Earth"), constants $1, \pi, r$ (where r is translated as a particular rock), relational predicates p, c, n (translated in English as "is a planet, is a cannonball, is a number"), a binary relational predicate $\circ$ (whose syntax is $x \circ y$ and whose translation in English is "x revolves around the fixed body y"), a binary relational predicate f (where $f(x, y)$ is translated as "x falls freely under the influence of y"), a binary functional symbol d ("distance between the centers of"), a unary functional symbol a ("acceleration"), a unary functional symbol T (where $T(x, y)$ is translated as "period of revolution of x around y"), binary functional symbols $:, \times$ ("division, multiplication"), and all the standard connectives, quantifiers, and parentheses. Note that we have no predicates for mass and force; this is remarkable because it shows that the Newtonian revolution has a purely geometric content. Now we introduce a theory T in L via its special axioms. The special axioms are as follows. First one asks that distances are numbers:

$$\forall x \forall y (n(d(x, y)))$$

and the same for accelerations, and times of revolution. (Note that we view all physical quantities as measured in centimeters and seconds.) For numbers we ask that multiplication and division of numbers are numbers:

$$(n(x) \wedge n(y)) \to (n(x : y) \wedge n(x \times y))$$

and that the usual laws relating $:$ and $\times$ hold. Here are two:

$$\forall x (x : x = 1).$$
$$\forall x \forall y \forall z \forall u ((x : y = z : u) \leftrightarrow (x \times u = z \times y)).$$

Exercise: write down all these laws. We sometimes write $\frac{x}{y}, 1/x, xy, x^2, x^3, ...$ in the usual sense. The above is a "baby mathematics" and this is all mathematics we need. Next we introduce an axiom whose justification is in mathematics, indeed in calculus; here we ignore the justification and just take this as an axiom. The axiom gives a formula for the acceleration of a body revolving in a circle around a fixed body. (See the exercise after this example.) Here is the axiom:

A) $\forall x \forall y \left(\left(x \circ y \right) \to \left(a(x, y) = \frac{4\pi^2 d(x, y)}{T^2(x, y)} \right) \right).$

To this one adds the following "obvious" axioms

O1) $\forall x (c(x) \to d(x, E) = R),$
O2) $M \circ E,$
R) $c(r),$
K1) $\forall x (p(x) \to (x \circ S)),$

saying that the distance between cannonballs and the center of the Earth is the radius of the Earth; that the Moon revolves around the Earth; that the rock r is a cannonball; and that all planets revolve around the Sun. (The latter is Kepler's first law in an approximate form; the full Kepler's first law specifies the shape of orbits as ellipses, etc.) Now we consider the following sentences (NOT AXIOMS!):

G) $\forall x \forall y ((c(x) \wedge c(y)) \to (a(x, E) = a(y, E)))$,

K3) $\forall x \forall y \left((p(x) \wedge p(y)) \to \left(\frac{d^3(x,S)}{T^2(x,S)} = \frac{d^3(y,S)}{T^2(y,S)} \right) \right)$,

N) $\forall x \forall y \forall z \left((f(x, z) \wedge f(y, z)) \to \left(\frac{a(x,z)}{1/d^2(x,z)} = \frac{a(y,z)}{1/d^2(y,z)} \right) \right)$.

G represents Galileo's great empirical discovery that all cannonballs (by which we mean here terrestrial airborne objects with no self-propulsion) have the same acceleration towards the Earth. $K3$ is Kepler's third law which is his empirical great discovery that the cubes of distances of planets to the Sun are in the same proportion as the squares of their periods of revolution. Kepler's second law about equal areas being swept in equal times is somewhat hidden in axiom A above. N is Newton's law of gravitation saying that the accelerations of any two bodies moving freely towards a fixed body are in the same proportion as the inverses of the squares of the respective distances to the (center of the) fixed body. Newton's great invention is the creation of a binary predicate f (where $f(x, y)$ is translated in English as "x is in free fall with respect to y") equipped with the following axioms

F1) $\forall x(c(x) \to f(x, E))$

F2) $f(M, E)$

F3) $\forall x(p(x) \to f(x, S))$

expressing the idea that cannonballs and the Moon moving relative to the Earth and planets moving relative to the Sun are instances of a more general predicate expressing "free falling." Finally let us consider the definition

$g = a(r, E)$

and the following sentence:

X) $g = \frac{4\pi^2 d^3(M,E)}{R^2 T^2(M,E)}$.

The main results are the following theorems in T:

THEOREM 11.20. $N \to X$.

Proof. See the exercise after this example.

THEOREM 11.21. $N \to G$.

Proof. See the exercise after this example.

THEOREM 11.22. $N \to K3$.

Proof. See the exercise after this example.

So if one accepts Newton's N then Galileo's G and Kepler's $K3$ follow, that is to say that N "unifies" terrestrial physics with the physics of Heaven. The beautiful thing is, however, that N not only unifies known paradigms but "predicts" new "facts," e.g., X. Indeed one can verify X using experimental (astronomical and terrestrial physics) data: if one enlarges our language to include numerals and numerical computations and if one introduces axioms as below (justified by measurements) then X becomes a theorem. Here are the additional axioms:

$g = 981$ (the number of centimeters per second squared representing g).

$\pi = \frac{314}{100}$ (approximate value).

R =number of centimeters representing the radius of the Earth (measured for the first time by Eratosthenes using shadows at two points on Earth).

$d(M,E)$ =number of centimeters representing the distance from Earth to Moon (measured using parallaxes).

$T(M,E)$ =number of seconds representing the time of revolution of the Moon (the equivalent of 28 days).

The fact that X is verified with the above data is the miraculous computation done by Newton that convinced him of the validity of his theory; see the exercise after this example.

REMARK 11.23. This part of Newton's early work had a series of defects: it was based on the circular (as opposed to elliptical) orbits, it assumed the center of the Earth (rather than all the mass of the Earth) as responsible for the effect on the cannonballs, it addressed only revolution around a fixed body (which is not realistic in the case of the Moon, since, for instance, the Earth itself is moving), and did not explain the difference between the d^3/T^2 of planets around the Sun and the corresponding quantity for the Moon and cannonballs relative to the Earth. Straightening these and many other problems is part of the reason why Newton postponed publication of his early discoveries. The final theory of Newton involves the introduction of absolute space and time, mass, and forces. The natural way to develop it is within mathematics, as mathematical physics; this is essentially the way Newton himself presented his theory in published form. However, the above example suggests that the real breakthrough was not mathematical but at the level of (pre-mathematical) logic.

EXERCISE 11.24.

1) Justify Axiom A above using calculus or even Euclidean geometry plus the definition of acceleration in an introductory physics course.

2) Prove Theorems 11.20, 11.21, and 11.22.

3) Verify that with the numerical data for $g, \pi, R, d(M,E), T(M,E)$ available from astronomy (find the numbers in astronomy books) the sentence X becomes a theorem. This is Newton's fundamental computation.

Part 2

Mathematics

CHAPTER 12

ZFC

Mathematics is a particular theory T_{set} (called set theory) in a particular language L_{set} (called the language of set theory) with specific axioms called the ZFC axioms (the Zermelo-Fraenkel+Choice axioms). So T_{set} will be $T(ZFC)$. We will introduce all of this presently.

The origins of set theory are in the work of Cantor. Set theory as presented in what follows is *not* Cantor set theory but rather an axiomatic version of it due essentially to Zermelo and Fraenkel. The difference between Cantor's paradigm and the Zermelo-Fraenkel paradigm lies, essentially, in the ontological status of sets. In contrast to Cantor, for whom sets were real collections of "things," any set in the definitions below is simply a constant (hence a single symbol) in a language.

METADEFINITION 12.1. The language L_{set} of set theory is the language with variables $x, y, z, ...$, constants $a, b, c, ..., A, B, C, ..., \mathcal{A}, \mathcal{B}, \mathcal{C}, ..., \alpha, \beta, \gamma, ...$, no functional symbol, a binary relational predicate $\in$, connectives $\vee, \wedge, \neg, \rightarrow, \leftrightarrow$, quantifiers $\forall, \exists$, equality $=$, and separators $(,), ,$. We also assume a witness assignment on L_{set} and definitions in L_{set} are given. We take the liberty to add, whenever it is convenient, new constants and new predicates to L_{set} together with definitions for each of these new symbols. The constants of L_{set} will be called sets.

REMARK 12.2. In the above metadefinition the witness assignment and the definitions were left unspecified. If one wants to pin down this concept further one can start with the language L_0 that has variables $x, y, z, ...$, no constants, no functional symbol, a binary relational predicate $\in$, connectives $\vee, \wedge, \neg, \rightarrow, \leftrightarrow$, quantifiers $\forall, \exists$, equality $=$, and separators (parentheses $(,)$ and comma); one assumes no definition in L_0. Then one can take L' to be the witness closure of L_0. Next we let L be obtained from L' by adding new constants $a, b, ...$, together with definitions for them, and also adding new relational predicates $\neq, \notin, \subset, \not\subset$ defined by

$\forall x \forall y ((x \neq y) \leftrightarrow (\neg(x = y)))$.
$\forall x \forall y ((x \notin y) \leftrightarrow (\neg(x \in y)))$.
$\forall x \forall y ((x \subset y) \leftrightarrow (\forall z ((z \in x) \rightarrow (z \in y))))$.
$\forall x \forall y ((x \not\subset y) \leftrightarrow (\neg(x \subset y)))$.

This L is the prototypical example we have in mind for L_{set}. Even with other examples of L_{set} we will maintain the above definitions of $\neq, \notin, \subset, \not\subset$. We say that x is a subset of y if $x \subset y$.

REMARK 12.3. We recall the fact that L_{set} being an object language it does not make sense to say that a sentence in it (such as, for instance, $a \in b$) is true or false.

REMARK 12.4. Later we will introduce the concept of "countable" set and we will show that not all sets are countable. On the other hand in set theory there are always only "finitely many" sets (in the sense that one is using finitely many symbols) although their collection may be increased any time, if necessary. Let us say that such a collection of symbols is "metacountable." This seems to be a paradox which is referred to as the "Skolem paradox." Of course this is not going to be a paradox: "metacountable" and "countable" will be two different concepts. The word "metacountable" belongs to the metalanguage and can be translated in English in terms of arranging symbols on a piece of paper; whereas "*b is countable*" is a definition in set theory. We define "*b is countable* ↔ $C(b)$" where $C(x)$ is a certain formula with free variable x in the language of set theory.

REMARK 12.5. There is a standard translation of the language L_{set} of set theory into the English language as follows:
$a, b, \dots$ are translated as "the set a," "the set b,"...
$\in$ is translated as "*belongs to the set*" or as "*is an element of the set*"
$=$ is translated as "*equals*"
$\subset$ is translated as "*is a subset*" or as "*is contained in*"
$\forall$ is translated as "*for all sets*"
$\exists$ is translated as "*there exists a set*"
while the connectives are translated in the standard way.

REMARK 12.6. Once we have a translation of L_{set} into English we can speak of argot and translation of L_{set} into argot; this simplifies comprehension of mathematical texts considerably.

REMARK 12.7. The standard translation of the language of set theory into English (in the remark above) is standard only by convention. A perfectly good different translation is, for instance, the one in which
$a, b, \dots$ are translated as "*crocodile a*," "*crocodile b*,"...
$\in$ is translated as "*is dreamt by the crocodile*"
$=$ is translated as "*has the same taste*"
$\forall$ is translated as "*for all crocodiles*"
$\exists$ is translated as "*there exists a crocodile*"
One could read mathematical texts in this translation; admittedly the English text that would result from this translation would be somewhat strange.

REMARK 12.8. Note that mathematics uses other symbols as well such as

$$\leq, \circ, +, \times, \sum a_n, \mathbb{Z}, \mathbb{Q}, \mathbb{R}, \mathbb{C}, \mathbb{F}_p, \equiv, \lim a_n, \int f(x)dx, \frac{df}{dx}, \dots$$

These symbols will originally be all sets (hence constants) and will be introduced through appropriate definitions (like the earlier definition of an elephant); they will all be defined through the predicate $\in$. In particular in the language of sets, $+$ or $\times$ are NOT originally functional symbols; and $\leq$ is NOT originally a relational predicate. However we will later tacitly enlarge the language of set theory by adding predicates (usually still denoted by) $+$ or $\leq$ via appropriate definitions.

We next introduce the (specific) axioms of set theory.

AXIOM 12.9. (Singleton axiom)
$$\forall x \exists y ((x \in y) \wedge (\forall z ((z \in y) \to (z = x)))).$$

The translation in argot is that for any set x there is a set y whose only element is x.

AXIOM 12.10. (Unordered pair axiom)

$$\forall x \forall x' \exists y((x \in y) \wedge (x' \in y) \wedge ((z \in y) \to ((z = x) \vee (z = x')))).$$

In argot the translation is that for any two sets x, x' there is a set that only has them as elements.

AXIOM 12.11. (Separation axioms) For any formula $P(x)$ in the language of sets, having a free variable x, we introduce an axiom

$$\forall y \exists z \forall x((x \in z) \leftrightarrow ((x \in y) \wedge (P(x)))).$$

The translation in argot is that for any set y there is a set z whose elements are all the elements x of y such that $P(x)$.

AXIOM 12.12. (Extensionality axiom)

$$\forall u \forall v((u = v) \leftrightarrow \forall x((x \in u) \leftrightarrow (x \in v))).$$

The translation in argot is that two sets u and v are equal if and only if they have the same elements.

AXIOM 12.13. (Union axiom)

$$\forall w \exists u \forall x((x \in u) \leftrightarrow (\exists t((t \in w) \wedge (x \in t)))).$$

The translation in argot is that for any set w there exists a set u such that for any x we have that x is an element of u if and only if x is an element of one of the elements of w.

AXIOM 12.14. (Empty set axiom)

$$\exists x \forall y(y \notin x).$$

The translation in argot is that there exists a set that has no elements.

AXIOM 12.15. (Power set axiom)

$$\forall y \exists z \forall x((x \in z) \leftrightarrow (\forall u((u \in x) \to (u \in y)))).$$

The translation in argot is that for any set y there is a set z such that a set x is an element of z if and only if all elements of x are elements of y.

For simplicity the rest of the axioms will be formulated in argot only.

DEFINITION 12.16. Two sets are disjoint if they have no element in common. The elements of a set are pairwise disjoint if any two elements are disjoint.

AXIOM 12.17. (Axiom of choice) For any set w whose elements are pairwise disjoint sets there is a set that has exactly one element in common with each of the sets in w.

AXIOM 12.18. (Axiom of infinity) There exists a set x such that x contains some element u and such that for any $y \in x$ there exists $z \in x$ with the property that y is the only element of z. Intuitively this axiom guarantees the existence of "infinite" sets.

AXIOM 12.19. (Axiom of foundation) For any set x there exists $y \in x$ such that x and y are disjoint.

One finally adds a technical list of axioms (indexed by formulas $P(x, y, z)$) about the "images of maps with parameters z":

AXIOM 12.20. (Axiom of replacement) If for any z and any u we have that $P(x, y, z)$ "defines y as a function of $x \in u$" (i.e., for any $x \in u$ there exists a unique y such that $P(x, y, z)$) then for all z there is a set v which is the "image of this map" (i.e., v consists of all y's with the property that there is an $x \in u$ such that $P(x, y, z)$). Here x, z may be tuples of variables.

EXERCISE 12.21. Write the axioms of choice, infinity, foundation, and replacement in the language of sets.

METADEFINITION 12.22. All of the above axioms form the ZFC system of axioms (Zermelo-Fraenkel+Choice). Set theory T_{set} is the theory $T(ZFC)$ in L_{set} generated by the ZFC axioms. Unless otherwise specified all theorems in the rest of the course are understood to be theorems in T_{set}.

REMARK 12.23. If set theory is to have a meaning at all (even though we decided to ignore it) one needs to be able to give a metaproof of the metasentence:

Set theory is consistent.

But the above metasentence involves quantifiers so no metaproof for it can be given. The only argument in favor of using set theory seems to be the fact that nobody could prove so far a statement of the form $P \wedge \neg P$ in this theory. Once mathematics is introduced and mathematical logic is developed one can introduce the concept of formalized set theory; we will be able then to consider a sentence C in set theory whose translation in English will be:

Formalized set theory is formally consistent.

(See the last part of the course dedicated to mathematical logic for an explanation of this sentence.) Gödel proved that the sentence C viewed as a sentence in formalized set theory is not "formally provable" (in a sense that, again, will be explained in the last part of the course).

REMARK 12.24. Note the important fact that the axioms did not involve constants. In the next chapter we investigate the constants, i.e., the sets.

From this moment on all proofs in this course will be written in argot. Also, unless otherwise stated, all proofs required to be given in the exercises must be written in argot.

CHAPTER 13

Sets

We will start here our discussion of sets and prove our first theorems in set theory. Recall that we introduced mathematics/set theory as being a specific theory $T(ZFC)$ in the language L_{set}, which we called T_{set}, where ZFC is a list of axioms that was described in the last chapter.

Recall the following:

METADEFINITION 13.1. A set is a constant in the language of set theory. Sets will be denoted by $a, b, ..., A, B, ..., \mathcal{A}, \mathcal{B}, ..., \alpha, \beta, \gamma, ...$

In what follows all definitions will be definitions in the language L_{set} of sets. Sometimes definitions are given in argotic L_{set}. Recall that some definitions introducing new constants are also simply referred to as notation.

We start by defining a new constant $\emptyset$ (called the empty set) as being equal to the witness for the axiom $\exists x \forall y (y \notin x)$; in other words $\emptyset$ is defined by

DEFINITION 13.2. $\emptyset = c^{\forall y(y \notin x)}$.

Note that $\forall y(y \notin \emptyset)$ is a theorem. In argot we say that $\emptyset$ is the "unique" set that contains no element.

Next if a is a set we introduce a new constant $\{a\}$ defined to be the witness for the sentence $\exists y P$ where

$$P \quad \text{equals} \quad \text{``}(a \in y) \wedge (\forall z((z \in y) \to (z = a))).\text{''}$$

In other words $\{a\}$ is defined by

DEFINITION 13.3. $\{a\} = c^P = c^{(a \in y) \wedge (\forall z((z \in y) \to (z = a)))}$.

The sentence $\exists y P$ is a theorem (use the singleton axiom) so the following is a theorem:

$$(a \in \{a\}) \wedge (\forall z((z \in \{a\}) \to (z = a))).$$

We can say (and we will usually say, by abuse of terminology) that $\{a\}$ is "the unique" set containing a only among its elements; we will often use this kind of abuse of terminology. In particular $\{\{a\}\}$ denotes the set whose only element is the set $\{a\}$, etc. Similarly, for any two sets a, b with $a \neq b$ denote by $\{a, b\}$ the set that only has a and b as elements; the set $\{a, b\}$ is a witness for a theorem that follows from the unordered pair axiom. Also whenever we write $\{a, b\}$ we implicitly understand that $a \neq b$.

Next, for any set A and any formula $P(x)$ in the language of sets, having one free variable x we denote by $A(P)$ or $\{a \in A; P(a)\}$ or $\{x \in A; P(x)\}$ the set whose elements are the elements $a \in A$ such that $P(a)$; so the set $A(P)$ equals by definition the witness for the separation axiom that corresponds to A and P. More precisely we have:

DEFINITION 13.4. $A(P) = \{x \in A; P(x)\} = c^{\exists z \forall x((x \in z) \leftrightarrow (x \in A) \wedge P(x)))}$.

LEMMA 13.5. *If $A = \{a\}$ and $B = \{b, c\}$ then $A \neq B$.*

Proof. We proceed by contradiction. So assume $A = \{a\}$, $B = \{b, c\}$, and $A = B$ and seek a contradiction. Indeed since $a \in A$ and $A = B$, by the extensionality axiom we get $a \in B$. Hence $a = b$ or $a = c$. Assume $a = b$ and seek a contradiction. (In the same way we get a contradiction by assuming $a = c$.) Since $a = b$ we get $B = \{a, c\}$. Since $c \in B$ and $A = B$, by the extensionality axiom we get $c \in A$. So $c = a$. Since $a = b$ we get $b = c$. But by our notation for sets (elements listed are distinct) we have $b \neq c$, a contradiction.   □

EXERCISE 13.6. Prove that:
1) If $\{a\} = \{b\}$ then $a = b$.
2) $\{a, b\} = \{b, a\}$.
3) $\{a\} = \{x \in \{a, b\}; x \neq b\}$.
4) There is a set b whose only elements are $\{a\}$ and $\{a, \{a\}\}$; so

$$b = \{\{a\}, \{a, \{a\}\}\}.$$

To make our definitions (notation) more reader friendly we will begin to express them in metalanguage as in the following example.

DEFINITION 13.7. For any two sets A and B we define the set $A \cup B$ (called the union of A and B) as the set such that for all c, $c \in A \cup B$ if and only if $c \in A$ or $c \in B$; the set $A \cup B$ is a witness for the union axiom.

EXERCISE 13.8. Explain in detail the definition of $A \cup B$ using a witness notation as in 13.2, 13.3, 13.4.

DEFINITION 13.9. The difference between the set A and the set B is the set

$$A \backslash B = \{c \in A; c \notin B\}.$$

DEFINITION 13.10. The intersection of the sets A and B is the set

$$A \cap B = \{c \in A; c \in B\}.$$

EXERCISE 13.11. Prove that
1) $A(P \wedge Q) = A(P) \cap A(Q)$;
2) $A(P \vee Q) = A(P) \cup A(Q)$;
3) $A(\neg P) = A \backslash A(P)$.

EXERCISE 13.12. Prove that if a, b, c are such that $(a \neq b) \wedge (b \neq c) \wedge (a \neq c)$ then there is a set denoted by $\{a, b, c\}$ whose only elements are a, b, c; in other words prove the following sentence:

$$\forall x \forall x' \forall x'' \exists y ((x \in y) \wedge (x' \in y) \wedge (x'' \in y) \wedge (z \in y) \rightarrow ((z = x) \vee (z = x') \vee (z = x'')))$$

Hint: Use the singleton axiom, the unordered pair axiom, and the union axiom, applied to the set $\{\{a\}, \{b, c\}\}$.

DEFINITION 13.13. Similarly one defines sets $\{a, b, c, d\}$, etc. Whenever we write $\{a, b, c\}$ or $\{a, b, c, d\}$, etc., we imply that the elements in each set are pairwise unequal (pairwise distinct). Also denote by

$$\{a, b, c, ...\}$$

any set d such that
$$(a \in d) \wedge (b \in d) \wedge (c \in d)$$
So the dots indicate that there may be other elements in d other than a, b, c; also note that when we write $\{a, b, c, ...\}$ we implicitly imply that a, b, c are pairwise distinct.

EXERCISE 13.14.
1) Prove that $\{\emptyset\} \neq \emptyset$.
2) Prove that $\{\{\emptyset\}\} \neq \{\emptyset\}$.

EXERCISE 13.15. Prove that $A = B$ if and only if $A \subset B$ and $B \subset A$.

EXERCISE 13.16. Prove that:
1) $\{a, b, c\} = \{b, c, a\}$.
2) $\{a, b\} \neq \{a, b, c\}$. Hint: Use $c \neq a, c \neq b$.
3) $\{a, b, c\} = \{a, b, d\}$ if and only if $c = d$.

EXERCISE 13.17. Let $A = \{a, b, c\}$ and $B = \{c, d\}$. Prove that
1) $A \cup B = \{a, b, c, d\}$,
2) $A \cap B = \{c\}$, $A \backslash B = \{a, b\}$.

EXERCISE 13.18. Let $A = \{a, b, c, d, e, f\}$, $B = \{d, e, f, g, h\}$. Compute
1) $A \cap B$,
2) $A \cup B$,
3) $A \backslash B$,
4) $B \backslash A$,
5) $(A \backslash B) \cup (B \backslash A)$.

EXERCISE 13.19. Prove the following:
1) $A \cap B \subset A$,
2) $A \subset A \cup B$,
3) $A \cap (B \cup C) = (A \cap B) \cup (A \cap C)$,
4) $A \cup (B \cap C) = (A \cup B) \cap (A \cup C)$,
5) $(A \backslash B) \cap (B \backslash A) - \emptyset$.

DEFINITION 13.20. For any set A we define the set $\mathcal{P}(A)$ as the set whose elements are the subsets of A; we call $\mathcal{P}(A)$ the power set of A; $\mathcal{P}(A)$ is a witness for (a theorem obtained from) the power set axiom.

EXERCISE 13.21. Explain in detail the definition of $\mathcal{P}(A)$ (using the witness notation).

EXAMPLE 13.22. If $A = \{a, b, c\}$ then
$$\mathcal{P}(A) = \{\emptyset, \{a\}, \{b\}, \{c\}, \{a, b\}, \{a, c\}, \{b, c\}, \{a, b, c\}\}.$$

EXERCISE 13.23. Let $A = \{a, b, c, d\}$. Write down the set $\mathcal{P}(A)$.

EXERCISE 13.24. Let $A = \{a, b\}$. Write down the set $\mathcal{P}(\mathcal{P}(A))$.

DEFINITION 13.25. (Ordered pairs) Let A and B be sets and let $a \in A$, $b \in B$. If $a \neq b$ the ordered pair (a, b) is the set $\{\{a\}, \{a, b\}\}$. We sometimes say "pair" instead of "ordered pair." If $a = b$ the pair $(a, b) = (a, a)$ is the set $\{\{a\}\}$. Note that $(a, b) \in \mathcal{P}(\mathcal{P}(A \cup B))$.

DEFINITION 13.26. For any sets A and B we define the product of A and B as the set

$$A \times B = \{c \in \mathcal{P}(\mathcal{P}(A \cup B)); \exists x \exists y ((x \in A) \wedge (y \in B) \wedge (c = (a, b)))\}.$$

This is the set whose elements are exactly the pairs (a, b) with $a \in A$ and $b \in B$.

PROPOSITION 13.27. $(a, b) = (c, d)$ *if and only if* $a = c$ *and* $b = d$.

Proof. We need to prove that
1) If $a = c$ and $b = d$ then $(a, b) = (c, d)$ and
2) If $(a, b) = (c, d)$ then $a = c$ and $b = d$.
Now 1) is obvious. To prove 2) assume $(a, b) = (c, d)$.
Assume first $a \neq b$ and $c \neq d$. Then by the definition of pairs we know that

$$\{\{a\}, \{a, b\}\} = \{\{c\}, \{c, d\}\}.$$

Since $\{a\} \in \{\{a\}, \{a, b\}\}$ it follows (by the extensionality axiom) that $\{a\} \in \{\{c\}, \{c, d\}\}$. Hence either $\{a\} = \{c\}$ or $\{a\} = \{c, d\}$. But as seen before $\{a\} \neq \{c, d\}$. So $\{a\} = \{c\}$. Since $a \in \{a\}$ it follows that $a \in \{c\}$ hence $a = c$. Similarly since $\{a, b\} \in \{\{a\}, \{a, b\}\}$ we get $\{a, b\} \in \{\{c\}, \{c, d\}\}$. So either $\{a, b\} = \{c\}$ or $\{a, b\} = \{c, d\}$. Again as seen before $\{a, b\} \neq \{c\}$ so $\{a, b\} = \{c, d\}$. So $b \in \{c, d\}$. So $b = c$ or $b = d$. Since $a \neq b$ and $a = c$ we get $b \neq c$. Hence $b = d$ and we are done in case $a \neq b$ and $c \neq d$.

Assume next $a = b$ and $c = d$. Then by the definition of pairs in this case we have $\{\{a\}\} = \{\{c\}\}$ and as before this implies $\{a\} = \{c\}$ hence $a = c$ so we are done in this case as well.

Finally assume $a = b$ and $c \neq d$. (The case $a \neq b$ and $c = d$ is treated similarly.) By the definition of pairs we get

$$\{\{a\}\} = \{\{c\}, \{c, d\}\}.$$

We get $\{c, d\} \in \{\{a\}\}$. Hence $\{c, d\} = \{a\}$ which is impossible, as seen before. This ends the proof.   □

EXERCISE 13.28. If $A = \{a, b, c\}$ and $B = \{c, d\}$ then

$$A \times B = \{(a, c), (a, d), (b, c), (b, d), (c, c), (c, d)\}.$$

Hint: By the above Proposition the pairs are distinct.

EXERCISE 13.29. Prove that
1) $(A \cap B) \times C = (A \times C) \cap (B \times C)$,
2) $(A \cup B) \times C = (A \times C) \cup (B \times C)$.

EXERCISE 13.30. Prove that

$$\neg (\exists y \forall z (z \in y))$$

In argot this says that there does not exist a set T such that for any set A we have $A \in T$. (Intuitively there is no set such that all sets belong to it.)

Hint: Assume there is such a T and derive a contradiction. To derive a contradiction consider the set

$$S = \{x \in T; x \notin x\}.$$

There are two possibilities. First $S \notin S$. Since $S \in T$, we get that $S \in S$, a contradiction. The second possibility is that $S \in S$. Since $S \in T$, we get that $S \notin S$, which is again a contradiction.

REMARK 13.31. Before the advent of ZFC Russell showed, using the set S above, that Cantor's set theory leads to a contradiction; this is referred to as the "Russell paradox." Within ZFC Russell's paradox, in its original form, disappears. Whether there are other forms of this paradox, or similar paradoxes, that survive in ZFC it is not clear.

CHAPTER 14

Maps

The concept of map (or function) has a long history. Originally functions were understood to be given by more or less explicit "formulae" (polynomial, rational, algebraic, and later by series). Controversies around what the "most general" functions should be arose, for instance, in connection with solving partial differential equations (by means of trigonometric series); this is somewhat parallel to the controversy around what the "most general" numbers should be that arose in connection with solving algebraic equations (such as $x^2 = 2$, $x^2 = -1$, or higher degree equations with no solutions expressed by radicals, etc.). The notion of "completely arbitrary" function gradually arose through the work of Dirichlet, Riemann, Weierstrass, Cantor, etc. Here is the definition:

DEFINITION 14.1. A map (or function) from a set A to a set B is a subset $F \subset A \times B$ such that for any $a \in A$ there is a unique $b \in B$ with $(a, b) \in F$. If $(a, b) \in F$ we write $F(a) = b$ or $a \mapsto b$ or $a \mapsto F(a)$. We also write $F : A \to B$ or $A \xrightarrow{F} B$.

REMARK 14.2. The above defines a new (ternary) relational predicate μ equal to "...*is a map from* ... *to* ...." Also we may introduce a new functional symbol $\widehat{F}$ by

$$\forall x \forall y ((\mu(F, A, B) \wedge (x \in A) \wedge (y \in B)) \to ((\widehat{F}(x) = y) \leftrightarrow ((x, y) \in F))).$$

Here F, A, B can be constants or variables depending on the context. We will usually drop the ^ (or think of the argot translation as dropping the hat). Also note that what we call a map $F \subset A \times B$ corresponds to what in elementary mathematics is called the graph of a map.

EXAMPLE 14.3. The set

(14.1) $$F = \{(a, a), (b, c)\} \subset \{a, b\} \times \{a, b, c\}$$

is a map and $F(a) = a$, $F(b) = c$. On the other hand the subset

$$F = \{(a, b), (a, c)\} \subset \{a, b\} \times \{a, b, c\}$$

is not a map.

DEFINITION 14.4. For any A the identity map $I : A \to A$ is defined as $I(a) = a$, i.e.,

$$I = I_A = \{(a, a); a \in A\} \subset A \times A.$$

DEFINITION 14.5. A map $F : A \to B$ is injective (or an injection, or one-to-one) if $F(a) = F(c)$ implies $a = c$.

DEFINITION 14.6. A map $F : A \to B$ is surjective (or a surjection, or onto) if for any $b \in B$ there exists an $a \in A$ such that $F(a) = b$.

EXAMPLE 14.7. The map (14.1) is injective and not surjective.

EXERCISE 14.8. Give an example of a map which is surjective and not injective.

EXERCISE 14.9. Let $A \subset B$. Prove that there is an injective map $i : A \to B$ such that $i(a) = a$ for all $a \in A$. We call i the inclusion map; we sometimes say $A \subset B$ is the inclusion map.

EXERCISE 14.10. (Composition) Prove that if $F : A \to B$ and $G : B \to C$ are two maps then there exists a unique map $H : A \to C$ such that $H(a) = G(F(a))$ for all a. We write $H = G \circ F$ and call the latter the composition of G with F. Hint: We let $(a, c) \in H$ if and only if there exists $b \in B$ with $(a, b) \in F$, $(b, c) \in G$.

DEFINITION 14.11. (Restriction) If $F : A \to B$ is a map and $A' \subset A$ then the composition of F with the inclusion map $A' \subset A$ is called the restriction of F to A' and is denoted by $F_{|A'} : A' \to B$.

DEFINITION 14.12. (Commutative diagram) By a commutative diagram of sets

$$
\begin{array}{ccc}
A & \overset{F}{\to} & B \\
U \downarrow & & \downarrow V \\
C & \overset{G}{\to} & D
\end{array}
$$

we mean a collection of sets and maps as above with the property that $G \circ U = V \circ F$.

EXERCISE 14.13. Prove that if $F \circ G$ is surjective then F is surjective. Prove that if $F \circ G$ is injective then G is injective.

EXERCISE 14.14. Prove that the composition of two injective maps is injective and the composition of two surjective maps is surjective.

DEFINITION 14.15. A map is bijective (or a bijection) if it is injective and surjective.

Here is a fundamental theorem in set theory:

THEOREM 14.16. (Bernstein's Theorem) If A and B are sets and if there exist injective maps $F : A \to B$ and $G : B \to A$ then there exists a bijective map $H : A \to B$.

The reader may attempt to prove this after he/she gets to the chapter on induction.

EXERCISE 14.17. Prove that if $F : A \to B$ is bijective then there exists a unique bijective map denoted by $F^{-1} : B \to A$ such that $F \circ F^{-1} = I_B$ and $F^{-1} \circ F = I_A$. F^{-1} is called the inverse of F.

EXERCISE 14.18. Let $F : \{a, b, c\} \to \{c, d, e\}$, $F(a) = d$, $F(b) = c$, $F(c) = e$. Prove that F has an inverse and compute F^{-1}.

EXERCISE 14.19. Prove that if A and B are sets then there exist maps $F : A \times B \to A$ and $G : A \times B \to B$ such that $F(a, b) = a$ and $G(a, b) = b$ for all $(a, b) \in A \times B$. (These are called the first and the second projection.) Hint: For G show that $G = \{((a, b), c); c = b\} \subset (A \times B) \times B$ is a map.

EXERCISE 14.20. Prove that $(A \times B) \times C \to A \times (B \times C)$, $((a, b), c) \mapsto (a, (b, c))$ is a bijection.

DEFINITION 14.21. Write $A \times B \times C$ instead of $(A \times B) \times C$ and write (a, b, c) instead of $((a, b), c)$. We call (a, b, c) a triple. Write $A^2 = A \times A$ and $A^3 = A \times A \times A$. More generally adopt this notation for arbitrary number of factors. Elements like $(a, b), (a, b, c), (a, b, c, d)$, etc. will be called tuples.

THEOREM 14.22. *If A is a set then there is no bijection between A and $\mathcal{P}(A)$*

Proof. Assume there exists a bijection $F : A \to \mathcal{P}(A)$ and seek a contradiction. Consider the set

$$B = \{a \in A; a \notin F(a)\} \in \mathcal{P}(A).$$

Since F is subjective there exists $b \in A$ such that $B = F(b)$. There are two cases: either $b \in B$ or $b \notin B$. If $b \in B$ then $b \in F(b)$ so $b \notin B$, a contradiction. If $b \notin B$ then $b \notin F(b)$ so $b \in B$, a contradiction, and we are done. $\square$

REMARK 14.23. Note the similarity between the above argument and the argument showing that there is no set having all sets as elements (the "Russell paradox").

DEFINITION 14.24. Let S be a set of sets and I a set. A family of sets in S indexed by I is a map $I \to S$, $i \mapsto A_i$. We sometimes drop the reference to S. We also write $(A_i)_{i \in I}$ to denote this family. By the union axiom for any such family there is a set (denoted by $\bigcup_{i \in I} A_i$, called their union) such that for all x we have that $x \in \bigcup_{i \in I} A_i$ if and only if there exists $i \in I$ such that $x \in A_i$. Also a set (denoted by $\bigcap_{i \in I} A_i$, called their intersection) exists such that for all x we have that $x \in \bigcap_{i \in I} A_i$ if and only if for all $i \in I$ we have $x \in A_i$. A family of elements in $(A_i)_{i \in I}$ is a map $I \to \bigcup_{i \in I} A_i$, $i \mapsto a_i$, such that for all $i \in I$ we have $a_i \in A_i$. Such a family of elements is denoted by $(a_i)_{i \in I}$. One defines the product $\prod_{i \in I} A_i$ as the set of all families of elements $(a_i)_{i \in I}$.

EXERCISE 14.25. Check that for $I = \{i, j\}$ the above definitions of $\cup, \cap, \prod$ yield the usual definition of $A_i \cap A_j$, $A_i \cap A_j$, and $A_i \times A_j$.

DEFINITION 14.26. Let $F : A \to B$ be a map and $X \subset A$. Define the image of X as the set

$$F(X) = \{y \in B; \exists x \in X, \ y = F(x)\} \subset B.$$

If $Y \subset B$ define the inverse image (or preimage) of Y as the set

$$F^{-1}(Y) = \{x \in A; F(x) \in Y\} \subset A.$$

For $y \in B$ define

$$F^{-1}(y) = \{x \in A; F(x) = y\}.$$

(Note that $F^{-1}(Y)$, $F^{-1}(y)$ are defined even if the inverse map F^{-1} does not exist, i.e., even if F is not bijective.)

EXERCISE 14.27. Let $F : \{a, b, c, d, e, f, g\} \to \{c, d, e, h\}$, $F(a) = d$, $F(b) = c$, $F(c) = e$, $F(d) = c$, $F(e) = d$, $F(f) = c$, $F(g) = c$. Let $X = \{a, b, c\}$, $Y = \{c, h\}$. Compute $F(X)$, $F^{-1}(Y)$, $F^{-1}(c)$, $F^{-1}(h)$.

EXERCISE 14.28. Prove that if $F : A \to B$ is a map and $X \subset X' \subset A$ are subsets then $F(X) \subset F(X')$.

EXERCISE 14.29. Prove that if $F : A \to B$ is a map and $(X_i)_{i \in I}$ is a family of subsets of A then

$$F(\cup_{i \in I} X_i) = \cup_{i \in I} F(X_i),$$
$$F(\cap_{i \in I} X_i) \subset \cap_{i \in I} F(X_i).$$

If in addition F is injective show that

$$F(\cap_{i \in I} X_i) = \cap_{i \in I} F(X_i).$$

Give an example showing that the latter may fail if F is not injective.

EXERCISE 14.30. Prove that if $F : A \to B$ is a map and $Y \subset Y' \subset B$ are subsets then $F^{-1}(Y) \subset F^{-1}(Y')$.

EXERCISE 14.31. Prove that if $F : A \to B$ is a map and $(Y_i)_{i \in I}$ is a family of subsets of B then

$$F^{-1}(\cup_{i \in I} Y_i) = \cup_{i \in I} F^{-1}(Y_i),$$
$$F^{-1}(\cap_{i \in I} Y_i) = \cap_{i \in I} F^{-1}(Y_i).$$

(So here one does not need injectivity like in the case of unions.)

DEFINITION 14.32. If A and B are sets we denote by $B^A \subset \mathcal{P}(A \times B)$ the set of all maps $F : A \to B$; sometimes one writes $Map(A, B) = B^A$.

EXERCISE 14.33. Let $0, 1$ be two elements. Prove that the map $\{0, 1\}^A \to \mathcal{P}(A)$ sending $F : A \to \{0, 1\}$ into $F^{-1}(1) \in \mathcal{P}(A)$ is a bijection.

EXERCISE 14.34. Find a bijection between $(C^B)^A$ and $C^{A \times B}$. Hint: Send $F \in (C^B)^A$, $F : A \to C^B$, into the set (map)

$$\{((a, b), c) \in (A \times B) \times C; (b, c) \in F(a)\}.$$

CHAPTER 15

Relations

A basic notion in set theory is that of relation; we shall investigate in some detail two special cases: order relations and equivalence relations.

DEFINITION 15.1. If A is a set then a relation on A is a subset $R \subset A \times A$. If $(a, b) \in R$ we write aRb.

REMARK 15.2. Exactly as in Remark 14.2 the above defines a new (binary) relational predicate "... *is a relation on* ..." and we may introduce a corresponding new relational predicate (still denoted by R).

DEFINITION 15.3. A relation R is called an order if (writing $a \leq b$ instead of aRb) we have, for all $a, b, c \in A$, that
1) $a \leq a$ (reflexivity),
2) $a \leq b$ and $b \leq c$ imply $a \leq c$ (transitivity),
3) $a \leq b$ and $b \leq a$ imply $a = b$ (antisymmetry).

DEFINITION 15.4. One writes $a < b$ if $a \leq b$ and $a \neq b$.

DEFINITION 15.5. An order relation is called a total order if for any $a, b \in A$ either $a \leq b$ or $b \leq a$. Alternatively we say A is totally ordered (by $\leq$).

EXAMPLE 15.6. For instance if $A = \{a, b, c, d\}$ then
$$R = \{(a, a), (b, b), (c, c), (d, d), (a, b), (b, c), (a, c)\}$$
is an order but not a total order.

EXERCISE 15.7. Let $R_0 \subset A \times A$ be a relation and assume R_0 is contained in an order relation $R_1 \subset A \times A$. Let
$$R = \bigcap_{R' \supset R_0} R'$$
be the intersection of all order relations R' containing R_0. Prove that R is an order relation and it is the smallest order relation containing R_0 in the sense that it is contained in any order relation that contains R_0.

EXERCISE 15.8. Let $A = \{a, b, c, d, e\}$ and $R_0 = \{(a, b), (b, c), (c, d), (c, e)\}$. Find an order relation containing R_0. Find the smallest order relation R containing R_0. Show that R is not a total order.

EXERCISE 15.9. Let A be a set. For any subsets $X \subset A$ and $Y \subset A$ write $X \leq Y$ if and only if $X \subset Y$. This defines a relation on the set $\mathcal{P}(A)$. Prove that this is an order relation. Give an example showing that this is not in general a total order.

DEFINITION 15.10. An ordered set is a pair $(A, \leq)$ where A is a set and $\leq$ is an order relation on A.

DEFINITION 15.11. Let $(A, \leq)$ and $(A', \leq')$ be ordered sets. A map $F : A \to A'$ is called increasing if for any $a, b \in A$ with $a \leq b$ we have $F(a) \leq' F(b)$.

EXERCISE 15.12. Prove that if $(A, \leq)$, $(A', \leq')$, $(A'', \leq'')$ are ordered sets and $G : A \to A'$, $F : A' \to A''$ are increasing then $F \circ G : A \to A''$ is increasing.

DEFINITION 15.13. Let A be a set with an order $\leq$. We say $\alpha \in A$ is a minimal element of A if for all $a \in A$ such that $a \leq \alpha$ we must have $a = \alpha$.

DEFINITION 15.14. Let A be a set with an order $\leq$. We say $\beta \in A$ is a maximal element of A if for all $b \in A$ such that $\beta \leq b$ we must have $\beta = b$.

DEFINITION 15.15. Let A be a set with an order $\leq$ and let $B \subset A$. We say $m \in B$ is a minimum element of B if for all $b \in B$ we have $m \leq b$. If a minimum element exists it is unique (check!) and we denote it by min B. Note that if min B exists then, by definition, min B belongs to B.

DEFINITION 15.16. Let A be a set with an order $\leq$ and $B \subset A$. We say $M \in B$ is a maximum element of B if for all $b \in B$ we have $b \leq M$. If a maximum element exists it is unique and we denote it by max B. Again, if max B exists then by definition it belongs to B.

DEFINITION 15.17. Let A be a set with an order $\leq$ and let $B \subset A$. An element $u \in A$ is called an upper bound for B if $b \leq u$ for all $b \in B$. We also say that B is bounded from above by b. An element $l \in A$ is called a lower bound for B if $l \leq b$ for all $b \in B$; we also say B is bounded from below by l. If the set of upper bounds of B has a minimum element we call it the supremum of B and we denote it by sup B; if the set of lower bounds of B has a maximum element we call it the infimum of B and we denote it by inf B. (Note that if one of sup B and inf B exists that element is by definition in A but does not necessarily belong to B.) We say B is bounded if it has both an upper bound and a lower bound.

EXERCISE 15.18. Consider the set A and the order $\leq$ defined by the relation R in Exercise 40.7. Does A have a maximum element? Does A have a minimum element? Are there maximal elements in A? Are there minimal elements in A? List all these elements in case they exist. Let $B = \{b, c\}$. Is B bounded? List all the upper bounds of B. List all the lower bounds of B. Does the supremum of B exist? If yes does it belong to B? Does the infimum of B exist? Does it belong to B?

DEFINITION 15.19. A well ordered set is an ordered set $(A, \leq)$ such that any non-empty subset $B \subset A$ has a minimum element.

EXERCISE 15.20. Prove that any well ordered set is totally ordered.

REMARK 15.21. Later, when we will have introduced the ordered set of integers and the ordered set of rational numbers we will see that the non-negative integers are well ordered but the non-negative rationals are not well ordered.

The following theorems can be proved (but their proof is beyond the scope of this course):

THEOREM 15.22. *(Zorn's lemma) Assume $(A, \leq)$ is an ordered set. Assume that any non-empty totally ordered subset $B \subset A$ has an upper bound in A. Then A has a maximal element.*

THEOREM 15.23. *(Well ordering principle) Let A be a set. Then there exists an order relation $\leq$ on A such that $(A, \leq)$ is well ordered.*

REMARK 15.24. It can be proved that if one removes from the axioms of set theory the axiom of choice then the axiom of choice, Zorn's lemma, and the well ordering principle are all equivalent.

EXERCISE 15.25. Let $(A, \leq)$ and $(B, \leq)$ be totally ordered sets. Define a relation $\leq$ on $A \times B$ by

$$((a,b) \leq (a',b')) \leftrightarrow ((a \leq a') \vee ((a = a') \wedge (b \leq b'))).$$

Prove that $\leq$ is an order on $A \times B$ (it is called the lexicographic order) and that $(A \times B, \leq)$ is totally ordered. (Explain how this order is being used to order words in a dictionary.)

DEFINITION 15.26. A relation R is called an equivalence relation if (writing $a \sim b$ instead of aRb) we have, for all $a, b, c \in A$, that
   1) $a \sim a$ (reflexivity),
   2) $a \sim b$ and $b \sim c$ imply $a \sim c$ (transitivity),
   3) $a \sim b$ implies $b \sim a$ (symmetry);
we also say that $\sim$ is an equivalence relation.

EXERCISE 15.27. Let $R_0 \subset A \times A$ be a relation and let

$$R = \bigcap_{R' \supset R_0} R'$$

be the intersection of all equivalence relations R' containing R_0. Prove that R is an equivalence relation and it is the smallest equivalence relation containing R_0 in the sense that it is contained in any other equivalence relation that contains R_0.

DEFINITION 15.28. Given an equivalence relation $\sim$ as above for any $a \in A$ we may consider the set

$$\widehat{a} = \{c \in A; c \sim a\}$$

called the equivalence class of a.

DEFINITION 15.29. Sometimes, instead of $\widehat{a}$, one writes $\overline{a}$ or $[a]$.

EXERCISE 15.30. Prove that $\widehat{a} = \widehat{b}$ if and only if $a \sim b$.

EXERCISE 15.31. Prove that:
   1) if $\widehat{a} \cap \widehat{b} \neq \emptyset$ then $\widehat{a} = \widehat{b}$;
   2) $A = \bigcup_{a \in A} \widehat{a}$.

DEFINITION 15.32. If A is a set a partition of A is a family $(A_i)_{i \in I}$ if subsets $A_i \subset A$ such that:
   1) if $i \neq j$ then $A_i \cap A_j = \emptyset$
   2) $A = \bigcup_{i \in I} A_i$.

EXERCISE 15.33. Let A be a set and $\sim$ an equivalence relation on it. Prove that:

1) There exists a subset $B \subset A$ which contains exactly one element of each equivalence class (such a set is called a system of representatives. Hint: Use the axiom of choice).

2) The family $(\widehat{b})_{b \in B}$ is a partition of A.

EXERCISE 15.34. Let A be a set and $(A_i)_{i \in I}$ a partition of A. Define a relation R on A as follows:

$$R = \{(a, b) \in A \times A; \exists i((i \in I) \wedge (a \in A_i) \wedge (b \in A_i))\}.$$

Prove that R is an equivalence relation.

EXERCISE 15.35. Let A be a set. Prove that there is a bijection between the set of equivalence relations on A and the set of partitions of A. Hint: Use the above two exercises.

DEFINITION 15.36. The set of equivalence classes

$$\{\alpha \in \mathcal{P}(A); \exists a((a \in A) \wedge (\alpha = \widehat{a}))\}$$

is denoted by $A/\sim$ and is called the quotient of A by the relation $\sim$.

EXAMPLE 15.37. For instance if $A = \{a, b, c\}$ and

$$R = \{(a, a), (b, b), (c, c), (a, b), (b, a)\}$$

then R is an equivalence relation, $\widehat{a} = \widehat{b} = \{a, b\}$, $\widehat{c} = \{c\}$, and $A/\sim = \{\{a, b\}, \{c\}\}$.

EXERCISE 15.38. Let $A = \{a, b, c, d, e, f\}$ and $R_0 = \{(a, b), (b, c), (d, e)\}$. Find the smallest equivalence relation R containing R_0. Call it $\sim$. Write down the equivalence classes $\widehat{a}, \widehat{b}, \widehat{c}, \widehat{d}, \widehat{e}, \widehat{f}$. Write down the set $A/\sim$.

EXERCISE 15.39. Let S be a set. For any sets $X, Y \in S$ write $X \sim Y$ if and only if there exists a bijection $F : X \to Y$. This defines a relation on S. Prove that this is an equivalence relation.

EXERCISE 15.40. Let $S = \{A, B, C, D\}$, $A = \{a, b\}$, $B = \{b, c\}$, $C = \{x, y\}$, $D = \emptyset$. Let $\sim$ be the equivalence relation on S defined in the previous exercise. Write down the equivalence classes $\widehat{A}, \widehat{B}, \widehat{C}, \widehat{D}$ and write down the set $S/\sim$.

DEFINITION 15.41. An affine plane is a pair $(A, \mathcal{L})$ where A is a set and $\mathcal{L} \subset \mathcal{P}(A)$ is a set of subsets of A satisfying a series of properties (which we call, by abuse, axioms) which we now explain. It is convenient to introduce some terminology as follows. A is called the affine plane. The elements of A are called points. The elements L of $\mathcal{L}$ are called lines; so each such L is a subset of A. We say a point P lies on a line L if $P \in L$; we also say that L passes through P. We say that two lines intersect if they have a point in common; we say that two lines are parallel if they either coincide or they do not intersect. We say that 3 points are collinear if they lie on the same line. Here are the axioms that we impose:

1) There exist 3 points which are not collinear and any line has at least 2 points.

2) Any 2 distinct points lie on exactly one line.

3) If L is a line and P is a point not lying on L there exists exactly one line through P which is parallel to L.

REMARK 15.42. Note that we have not defined 2 or 3 yet; this will be done later when we introduce integers. The meaning of these axioms is, however, clearly expressible in terms that were already defined. For instance axiom 2 says that for any points P and Q with $P \neq Q$ there exists a line through P and Q; we do not need to define the symbol 2 to express this. The same holds for the use of the symbol 3.

EXERCISE 15.43. Prove that any two distinct non-parallel lines intersect in exactly one point.

EXERCISE 15.44. Let $A = \{a, b\} \times \{a, b\}$ and let $\mathcal{L} \subset \mathcal{P}(A)$ consist of all subsets of 2 elements; there are 6 of them. Prove that $(A, \mathcal{L})$ is an affine plane. (Again one can reformulate everything without reference to the symbols 2 or 6; one simply uses 2 or 6 letters and writes that they are pairwise unequal.)

EXERCISE 15.45. Let $A = \{a, b, c\} \times \{a, b, c\}$. Find all subsets $\mathcal{L} \subset \mathcal{P}(A)$ such that $(A, \mathcal{L})$ is an affine plane. (This is tedious !)

DEFINITION 15.46. A projective plane is a pair $(\overline{A}, \overline{\mathcal{L}})$ where $\overline{A}$ is a set and $\overline{\mathcal{L}} \subset \mathcal{P}(\overline{A})$ is a set of subsets of $\overline{A}$ satisfying a series of axioms which we now explain. Again it is convenient to introduce some terminology as follows. $\overline{A}$ is called the projective plane. The elements of $\overline{A}$ are called points, P. The elements $\overline{L}$ of $\overline{\mathcal{L}}$ are called lines; so each such $\overline{L} \subset \overline{A}$. We say a point P lies on a line $\overline{L}$ if $P \in \overline{L}$; we also say that $\overline{L}$ passes through P. We say that two lines intersect if they have a point in common; we say that two lines are parallel if they either coincide or they do not intersect. We say that 3 points are collinear if they lie on the same line. Here are the axioms that we impose:

1) There exist 3 points which are not collinear and any line has at least 3 points.
2) Any 2 distinct points lie on exactly one line.
3) Any 2 distinct lines meet in exactly one point.

EXAMPLE 15.47. One can attach to any affine plane $(A, \mathcal{L})$ a projective plane $(\overline{A}, \overline{\mathcal{L}})$ as follows. We introduce the relation $\parallel$ on $\mathcal{L}$ by letting $L \parallel L'$ if and only if L and L' are parallel. This is an equivalence relation (check!). Denote by $\widehat{L}$ the equivalence class of L. Then we consider the set of equivalence classes, $\overline{L}_\infty = \mathcal{L}/\parallel$; call this set the line at infinity. There exists a set $\overline{A}$ such that $\overline{A} = A \cup \overline{L}_\infty$ and $A \cap \overline{L}_\infty = \emptyset$. Define a line in $\overline{A}$ to be either $\overline{L}_\infty$ or set of the form $\overline{L} = L \cup \{\widehat{L}\}$. Finally define $\overline{\mathcal{L}}$ to be the set of all lines in $\overline{A}$.

EXERCISE 15.48. Explain why $\overline{A}$ exists. Check that $(\overline{A}, \overline{\mathcal{L}})$ is a projective plane.

EXERCISE 15.49. Describe the projective plane attached to the affine plane in Exercise 15.44; how many points does it have? How many lines?

CHAPTER 16

Operations

The concept of operation on a set is an abstraction of "familiar" operations such as addition and multiplication of numbers, composition of functions, etc. Sets with operations on them will be referred to as algebraic structures. The study of algebraic structures is referred to as (modern) algebra and took the shape known today through work (in number theory and algebraic geometry) done by Kronecker, Dedekind, Hilbert, Emmy Noether, etc. Here we introduce operations in general, and some algebraic structures such as rings, fields, and Boolean algebras. We prefer to postpone the introduction of other algebraic structures such as groups, vector spaces, etc., until more theory is being developed.

DEFINITION 16.1. A binary operation $\star$ on a set A is a map $\star : A \times A \to A$, $(a, b) \mapsto \star(a, b)$. We usually write $a \star b$ instead of $\star(a, b)$. For instance, we write $(a \star b) \star c$ instead of $\star(\star(a, b), c)$. Instead of $\star$ we sometimes use notation like $+, \times, \circ, \ldots$.

REMARK 16.2. Exactly as in Remark 14.2 the above defines a new (binary) relational predicate "... *is a binary operation on* ..." and we may introduce a corresponding new functional symbol (still denoted by $\star$).

DEFINITION 16.3. A unary operation $'$ on a set A is a map $' : A \to A$, $a \mapsto '(a)$. We usually write a' or $'a$ instead of $'(a)$. Instead of $'$ we sometimes use notation like $-, i, \ldots$.

EXAMPLE 16.4. Let $S = \{0, 1\}$ where $0, 1$ are two sets. Then there are 3 interesting binary operations on S denoted by $\wedge, \vee, +$ (and called supremum, infimum, and addition) defined as follows:

$$0 \wedge 0 = 0, \quad 0 \wedge 1 = 0, \quad 1 \wedge 0 = 0, \quad 1 \wedge 1 = 1;$$

$$0 \vee 0 = 0, \quad 0 \vee 1 = 1, \quad 1 \vee 0 = 1, \quad 1 \vee 1 = 1;$$

$$0 + 0 = 0, \quad 0 + 1 = 1, \quad 1 + 0 = 1, \quad 1 + 1 = 0.$$

The symbol $\wedge$ is also denoted by $\times$ or $\cdot$; it is referred to as multiplication. The symbol $+$ is also denoted by Δ. Also there is a unary operation $\neg$ on S defined by

$$\neg 1 = 0, \quad \neg 1 = 0.$$

Note that if we denote 0 and 1 by F and T then the operations $\wedge, \vee, \neg$ on $\{0, 1\}$ correspond exactly to the "logical operations" on F and T defined in the chapter on tautologies. This is not a coincidence!

EXERCISE 16.5. Compute $((0 \wedge 1) \vee 1) + (1 \wedge (0 \vee (1 + 1)))$.

DEFINITION 16.6. A Boolean algebra is a tuple

$$(A, \vee, \wedge, \neg, 0, 1)$$

where $\wedge, \vee$ are binary operations, $\neg$ is a unary operation, and $0, 1 \in A$ such that for all $a, b, c \in A$ the following "axioms" are satisfied:
1) $a \wedge (b \wedge c) = (a \wedge b) \wedge c$, $a \vee (b \vee c) = (a \vee b) \vee c$,
2) $a \wedge b = b \wedge a$, $a \vee b = b \vee a$,
3) $a \wedge 1 = a$, $a \vee 0 = a$,
4) $a \wedge (b \vee c) = (a \wedge b) \vee (a \wedge c)$, $a \vee (b \wedge c) = (a \vee b) \wedge (a \vee c)$
5) $a \wedge (\neg a) = 0$, $a \vee (\neg a) = 1$.

DEFINITION 16.7. A commutative unital ring (or simply a ring) is a tuple

$$(R, +, \times, -, 0, 1)$$

(sometimes referred to simply as R) where R is a set, $0, 1 \in R$, $+, \times$ are two binary operations (write $a \times b = ab$), and $-$ is a unary operation on R such that for any $a, b, c \in R$ the following hold:
1) $a + (b + c) = (a + b) + c$, $a + 0 = a$, $a + (-a) = 0$, $a + b = b + a$;
2) $a(bc) = a(bc)$, $1a = a$, $ab = ba$,
3) $a(b + c) = ab + ac$.

The element 1 is referred to as the identity; 0 is referred to as the zero element.

DEFINITION 16.8. We write $a + b + c$ instead of $(a + b) + c$ and abc for $(ab)c$. We write $a - b$ instead of $a + (-b)$.

DEFINITION 16.9. An element a of a ring R is invertible if there exists $a' \in R$ such that $aa' = 1$; this a' is then easily proved to be unique. It is called the inverse of a, and is denoted by a^{-1}. A ring R is called a field if $0 \neq 1$ and any non-zero element is invertible.

DEFINITION 16.10. A Boolean ring is a commutative unital ring such that $1 \neq 0$ and for all $a \in A$ we have $a^2 = a$.

EXERCISE 16.11. Prove that in a Boolean ring A we have $a + a = 0$ for all $a \in A$.

EXERCISE 16.12. Prove that
1) $(\{0, 1\}, \vee, \wedge, \neg, 0, 1)$ is a Boolean algebra.
2) $(\{0, 1\}, +, \times, I, 0, 1)$ is a Boolean ring and a field (I is the identity map).

EXERCISE 16.13. Prove that if a Boolean ring A is a field then $A = \{0, 1\}$.

DEFINITION 16.14. Let A be a set and let $S = \mathcal{P}(A)$ be the power set of A. Define the following operations on S:

$$\begin{aligned} X \wedge Y &= X \cap Y \\ X \vee Y &= X \cup Y \\ X \Delta Y &= (X \cup Y) \backslash (X \cap Y) \\ \neg X &= \complement X = A \backslash X. \end{aligned}$$

EXERCISE 16.15. Prove that
1) $(\mathcal{P}(A), \vee, \wedge, \neg, \emptyset, A)$ is a Boolean algebra;
2) $(\mathcal{P}(A), \Delta, \wedge, I, \emptyset, A)$ is a Boolean ring (I is the identity map).

Hint: For any $a \in A$ one can define a map $\psi_a : \mathcal{P}(A) \to \{0, 1\}$ by setting $\psi_a(X) = 1$ if and only if $a \in X$. Note that

1) $\psi_a(X \wedge Y) = \psi_a(X) \wedge \psi_a(Y)$,
2) $\psi_a(X \vee Y) = \psi_a(X) \vee \psi_a(Y)$,
3) $\psi_a(X \Delta Y) = \psi_a(X) + \psi_a(Y)$,
4) $\psi_a(\neg X) = \neg \psi_a(X)$.

Next note that $X = Y$ if and only if $\psi_a(X) = \psi_a(Y)$ for all $a \in A$. Use these functions to reduce the present exercise to Exercise 16.12.

DEFINITION 16.16. Given a subset $X \subset A$ one can define the characteristic function $\chi_X : A \to \{0,1\}$ by letting $\chi_X(a) = 1$ if and only if $a \in X$; in other words $\chi_X(a) = \psi_a(X)$.

EXERCISE 16.17. Prove that
1) $\chi_{X \vee Y}(a) = \chi_X(a) \vee \chi_Y(a)$,
2) $\chi_{X \wedge Y}(a) = \chi_X(a) \wedge \chi_Y(a)$,
3) $\chi_{X \Delta Y}(a) = \chi_X(a) + \chi_Y(a)$,
4) $\chi_{\neg X}(a) = \neg \chi_X(a)$.

DEFINITION 16.18. An algebraic structure is a tuple $(A, \star, \bullet, ..., \neg, -, ..., 0, 1, ...)$ where A is a set, $\star, \bullet, ...$ are binary operations, $\neg, -, ...$ are unary operations, and $1, 0, ...$ are given elements of A. (Some of these may be missing; for instance we could have only one binary operation, one given element, and no unary operations.) Assume we are given two algebraic structures

$$(A, \star, \bullet, ..., \neg, -, ..., 0, 1, ...) \quad \text{and} \quad (A', \star', \bullet', ..., \neg', -', ..., 0', 1', ...)$$

(with the same number of corresponding operations). A map $F : A \to A'$ is called a homomorphism if for all $a, b \in A$ we have:
1) $F(a \star b) = F(a) \star' F(b)$, $F(a \bullet b) = F(a) \bullet' F(b)$,...
2) $F(\neg a) = \neg' F(a)$, $F(-a) = -' F(a)$,...
3) $F(0) = 0'$, $F(1) = 1'$,....

EXAMPLE 16.19. A map $F : A \to A'$ between two commutative unital rings is called a homomorphism (of commutative unital rings) if for all $a, b \in A$ we have:
1) $F(a + b) = F(a) + F(b)$ and $F(ab) = F(a)F(b)$,
2) $F(-a) = -F(a)$ (prove that this is automatic !),
3) $F(0) = 0$ (prove that this is automatic !) and $F(1) = 1$.

EXERCISE 16.20. Prove that if $F : A \to A'$ is a homomorphism of algebraic structures and F is bijective then its inverse $F^{-1} : A' \to A$ is a homomorphism. Such an F will be called an isomorphism.

DEFINITION 16.21. A subset $\mathcal{A} \subset \mathcal{P}(A)$ is called a Boolean algebra of sets if the following hold:
1) $\emptyset \in \mathcal{A}$, $A \in \mathcal{A}$;
2) If $X, Y \in \mathcal{A}$ then $X \cap Y \in \mathcal{A}$, $X \cup Y \in \mathcal{A}$, $\mathcal{C}X \in \mathcal{A}$.
(Hence $(\mathcal{A}, \vee, \wedge, \mathcal{C}, \emptyset, A)$ is a Boolean algebra.)

EXERCISE 16.22. Prove that if $\mathcal{A}$ is a Boolean algebra of sets then for any $X, Y \in \mathcal{A}$ we have $X \Delta Y \in \mathcal{A}$. Prove that $(\mathcal{A}, \Delta, \cap, I, \emptyset, A)$ is a Boolean ring.

DEFINITION 16.23. A subset $\mathcal{B} \subset \mathcal{P}(A)$ is called a Boolean ring of sets if the following properties hold:
1) $\emptyset \in \mathcal{B}$, $A \in \mathcal{B}$;
2) If $X, Y \in \mathcal{B}$ then $X \cap Y \in \mathcal{B}$, $X \Delta Y \in \mathcal{B}$.
(Hence $(\mathcal{A}, \Delta, \vee, I, \emptyset, A)$ is a Boolean ring.)

EXERCISE 16.24. Prove that any Boolean ring of sets $\mathcal{B}$ is a Boolean algebra of sets.

DEFINITION 16.25. A commutative unital ordered ring (or simply an ordered ring) is a tuple
$$(R, +, \times, -, 0, 1, \leq)$$
where
$$(R, +, \times, -, 0, 1)$$
is a ring, $\leq$ is a total order on R, and for all $a, b, c \in R$ the following axioms are satisfied
1) If $a < b$ then $a + c < b + c$;
2) If $a < b$ and $c > 0$ then $ac < bc$.
We say that $a \in R$ is positive if $a > 0$; and that a is negative if $a < 0$. We say a is non-negative if $a \geq 0$.

EXERCISE 16.26. Prove that the ring $(\{0, 1\}, +, \times, -, 0, 1)$ has no structure of ordered ring i.e., there is no order $\leq$ on $\{0, 1\}$ such that $(\{0, 1\}, +, \times, -, 0, 1, \leq)$ is an ordered ring.

REMARK 16.27. We cannot give examples yet of ordered rings. Later we will see that the rings of integers, rationals, and reals have natural structures of ordered rings.

DEFINITION 16.28. Let R be an ordered ring and let $R_+ = \{a \in R; a \geq 0\}$. A finite measure space is a triple $(A, \mathcal{A}, \mu)$ where A is a set, $\mathcal{A} \subset \mathcal{P}(A)$ is a Boolean algebra of sets, and $\mu : \mathcal{A} \to R_+$ is a map satisfying the property that for any $X, Y \in \mathcal{A}$ with $X \cap Y = \emptyset$ we have
$$\mu(X \cup Y) = \mu(X) + \mu(Y).$$
If in addition $\mu(A) = 1$ we say $(A, \mathcal{A}, \mu)$ is a finite probability measure. We say that $X, Y \in \mathcal{A}$ are independent if $\mu(X \cap Y) = \mu(X) \cdot \mu(Y)$.

EXERCISE 16.29. Prove that in a finite measure space $\mu(\emptyset) = 0$ and for any $X, Y \in \mathcal{A}$ we have
$$\mu(X \cup Y) = \mu(X) + \mu(Y) - \mu(X \cap Y).$$

EXERCISE 16.30. Let $(A, \vee, \wedge, \neg, 0, 1)$ be a Boolean algebra. For any $a, b \in A$ set
$$a + b = (a \vee b) \wedge (\neg(a \wedge b)).$$
Prove that $(A, +, \wedge, I, 0, 1)$ is a Boolean ring (I the identity map).

EXERCISE 16.31. Let $(A, +, \times, -, 0, 1)$ be a Boolean ring. For any $a, b \in A$ let
$$\begin{aligned} a \vee b &= a + b - ab \\ a \wedge b &= ab \\ \neg a &= 1 - a. \end{aligned}$$
Prove that $(A, \vee, \wedge, \neg, 0, 1)$ is a Boolean algebra.

EXERCISE 16.32. Let X be a set and $(R, +, \cdot, -, 0, 1)$ a commutative unital ring. Let R^X be the set of all functions $X \to R$. For $F, G \in R^X$ we define $F + G, F \cdot G, -F, 0, 1 \in R^X$ by the formulae
$$(F + G)(x) = F(x) + G(x), \quad (F \cdot G)(x) = F(x) \cdot G(x),$$

$$(-F)(x) = -F(x), \quad 0(x) = 0, \quad 1(x) = x,$$

for all $x \in X$. The operations $F + G$ and $F \cdot G$ are called pointwise addition and multiplication of functions. Prove that

$$(R^X, +, \cdot, -, 0, 1)$$

is a commutative unital ring.

CHAPTER 17

Integers

In this Chapter we introduce the ring $\mathbb{Z}$ of integers and we prove some easy theorems about this concept.

DEFINITION 17.1. A well ordered ring is an ordered ring $(R, +, \times, 0, 1, \leq)$ with $1 \neq 0$ having the property that any non-empty subset of R which is bounded from below has a minimum element.

REMARK 17.2. If $(R, +, \times, 0, 1, \leq)$ is a well ordered ring then $(R, \leq)$ is not a priori a well ordered set. But if $R_{>0} = \{a \in R; a > 0\}$ then $(R_{>0}, \leq)$ is a well ordered set.

We have the following remarkable theorem in set theory T_{set}:

THEOREM 17.3. *There exists a well ordered ring.*

REMARK 17.4. The above theorem is formulated, as usual, in argot; but it should be understood as being a sentence Z in L_{set} of the form

$$\exists r \exists s \exists p \exists o \exists u \exists l (...)$$

where we take a variable r to stand for the ring, a variable s for the sum, p for the product, o for 0, u for 1, l for $\leq$, and the dots stand for the corresponding conditions in the definition of a well ordered ring, written in the language of sets. The sentence Z is complicated so we preferred to give the theorem not as a sentence in L_{set} but as a sentence in argot. This kind of abuse is very common.

REMARK 17.5. We are going to sketch the proof of Theorem 17.3 in an exercise below. The proof is involved. A cheap way to avoid the proof of this theorem is as follows: add this theorem to the ZFC axioms and let ZFC' be the resulting enriched system of axioms. Then replace T_{set} by the theory T'_{set} with axioms ZFC'. This is what all working mathematicians essentially do anyway.

DEFINITION 17.6. We let $\mathbb{Z}, +, \times, 0, 1, \leq$ be the witnesses for the sentence Z above; we call $\mathbb{Z}$ the ring of integers. In particular the conditions in the definition of rings (associativity, commutativity, etc.) and order (transitivity, etc.) become theorems for $\mathbb{Z}$. We also set $\mathbb{N} = \{a \in \mathbb{Z}; a > 0\}$ and we call $\mathbb{N}$ the set of natural numbers. Later we will prove the "essential uniqueness" of $\mathbb{Z}$.

REMARK 17.7. The only predicate in the language L_{set} of sets is $\in$ and the constants in this language are called sets. In particular when we consider the ordered ring of integers $(\mathbb{Z}, +, \times, 0, 1, \leq)$ the symbols $\mathbb{Z}, +, \times, 0, 1, \leq, \mathbb{N}$ are all constants (they are sets). In particular $+, \times$ are not originally functional symbols and $\leq$ is not originally a relational predicate. But, according to our conventions, we may introduce functional symbols (still denoted by $+, \times$) and a relational predicate (still denoted by $\leq$) via appropriate definitions. (This is because *"the set $+$ is a binary operation on $\mathbb{Z}$"* is a theorem, etc.)

EXERCISE 17.8. Prove that $1 \in \mathbb{N}$. Hint: Use $(-1) \times (-1) = 1$.

EXERCISE 17.9. Prove that if $a, b \in \mathbb{Z}$ and $ab = 0$ then either $a = 0$ or $b = 0$. Give an example of a ring where this is not true. (For the latter consider the Boolean ring $\mathcal{P}(A)$.)

EXERCISE 17.10. Prove that if $a \in \mathbb{Z}$ then the set $\{x \in \mathbb{Z}; a - 1 < x < a\}$ is empty. Hint: It is enough to show that $S = \{x \in \mathbb{Z}; 0 < x < 1\}$ is empty. Assume S is non-empty and let $m = \min S$. Then $0 < m^2 < m$, hence $0 < m^2 < 1$ and $m^2 < m$, a contradiction.

EXERCISE 17.11. Prove that if $a \in \mathbb{N}$ then $a = 1$ or $a - 1 \in \mathbb{N}$. Conclude that $\min \mathbb{N} = 1$. Hint: Use the previous exercise.

In what follows we sketch the main idea behind the proof of Theorem 17.3. We begin with the following:

DEFINITION 17.12. A Peano triple is a triple $(N, 1, \sigma)$ where N is a set, $1 \in N$, and $\sigma : N \to N$ is a map such that
  1) σ is injective;
  2) $\sigma(N) = N \backslash \{1\}$;
  3) for any subset $S \subset N$ if $1 \in S$ and $\sigma(S) \subset S$ then $S = N$.
The conditions 1, 2, 3 are called "Peano's axioms."

REMARK 17.13. Given a well ordered ring one can easily prove there exists a Peano triple; cf. Exercise 17.14 below. Conversely given a Peano triple one can prove there exists a well ordered ring; this is tedious and will be addressed in Exercise 17.15. The idea of proof of Theorem 17.3 is to prove in set theory (i.e., ZFC) that there exists a Peano triple; here the axiom of infinity is crucial. Then the existence of a well ordered ring follows.

EXERCISE 17.14. Prove that if Z is a well ordered ring, $N = \{x \in Z; x > 0\}$, and $\sigma : N \to N$ is $\sigma(x) = x + 1$ then $(N, 1, \sigma)$ is a Peano triple.

EXERCISE 17.15. This exercise gives some steps towards showing how to construct a well ordered ring from a given Peano triple. Assume $(N, 1, \sigma)$ is a Peano triple. For $y \in N$ let

$$A_y = \{\tau \in N^N; \tau(1) = \sigma(y), \forall x(\tau(\sigma(x)) = \sigma(\tau(x)))\}.$$

1) Prove that A_y has at most one element. Hint: If $\tau, \eta \in A_y$ and $S = \{x; \tau(x) = \eta(x)\}$ then $1 \in S$ and $\sigma(S) \subset S$; so $S = N$.

2) Prove that for any y, $A_y \neq \emptyset$. Hint: If $T = \{y \in N; A_y \neq \emptyset\}$ then $1 \in T$ and $\sigma(T) \subset T$; so $T = N$.

3) By 1 and 2 we may write $A_y = \{\tau_y\}$. Then define $+$ on N by $x + y = \tau_y(x)$.

4) Prove that $x + y = y + x$ and $(x + y) + z = x + (y + z)$ on N.

5) Prove that if $x, y \in N$, $x \neq y$, then there exists $z \in N$ such that either $y = x + z$ or $x = y + z$.

6) Define $N' = \{-\} \times N$, $Z = N' \cup \{0\} \cup N$ where 0 and $-$ are two sets. Naturally extend $+$ to Z.

7) Define $\times$ on N and then on Z in the same style as for $+$.

8) Define $\leq$ on N and prove $(N, \leq)$ is well ordered. Extend this to Z.

9) Prove that $(Z, +, \times, 0, 1, \leq)$ is a well ordered ring.

From now on we accept Theorem 17.3 (either as a theorem whose proof we summarily sketched or as an additional axiom for set theory).

DEFINITION 17.16. Define the natural numbers $2, 3, ..., 9$ by

$$
\begin{aligned}
2 &= 1 + 1 \\
3 &= 2 + 1 \\
&\cdots \\
9 &= 8 + 1.
\end{aligned}
$$

Define $10 = 2 \times 5$. Define $10^2 = 10 \times 10$, etc. Define symbols like 423 as being $4 \times 10^2 + 2 \times 10 + 3$, etc. This is called a decimal representation.

EXERCISE 17.17. Prove that $12 = 9 + 3$. Hint: We have:

$$
\begin{aligned}
12 &= 10 + 2 \\
&= 2 \times 5 + 2 \\
&= (1 + 1) \times 5 + 2 \\
&= 1 \times 5 + 1 \times 5 + 2 = 5 + 5 + 2 \\
&= 5 + 5 + 1 + 1 = 5 + 6 + 1 = 5 + 7 = 4 + 1 + 7 \\
&= 4 + 8 = 3 + 1 + 8 = 3 + 9 = 9 + 3.
\end{aligned}
$$

EXERCISE 17.18. Prove that $18 + 17 = 35$. Prove that $17 \times 3 = 51$.

REMARK 17.19. In Kant's analysis, statements like the ones in the previous exercise were viewed as synthetic; in contemporary mathematics, hence in the approach we follow, all these statements are, on the contrary, analytic statements. (The definition of analytic/synthetic is taken here in the sense of Leibniz and Kant.)

EXERCISE 17.20. Prove that $7 \leq 20$.

DEFINITION 17.21. For any integers $a, b \in \mathbb{Z}$ the set $\{x \in \mathbb{Z}; a \leq x \leq b\}$ will be denoted, for simplicity, by $\{a, ..., b\}$. This set is clearly empty if $a > b$. If other numbers in addition to a, b are specified then the meaning of our notation will be clear from the context; for instance $\{0, 1, ..., n\}$ means $\{0, ..., n\}$ whereas $\{2, 4, 6, ..., 2n\}$ will mean $\{2x; 1 \leq x \leq n\}$, etc. A similar convention applies if there are no numbers after the dots.

EXAMPLE 17.22. $\{-2, ..., 11\} = \{-2, -1, 0, 1, 2, 3, 4, 5, 6, 7, 8, 9, 10, 11\}$.

Recall that a subset $A \subset \mathbb{N}$ is bounded (equivalently bounded from above) if there exists $b \in \mathbb{N}$ such that $a \leq b$ for all $a \in A$; we say that A is bounded by b from above.

EXERCISE 17.23. Prove that $\mathbb{N}$ is not bounded.

EXERCISE 17.24. Prove that any subset of $\mathbb{Z}$ bounded from above has a maximum. Hint: If A is bounded from above by b consider the set $\{b - x; x \in A\}$.

DEFINITION 17.25. An integer a is even if there exists an integer b such that $a = 2b$. An integer is odd if it is not even.

EXERCISE 17.26. Prove that if a is odd then $a - 1$ is even. Hint: Consider the set $\{b \in \mathbb{N}; 2b \geq a\}$, and let c be the minimum element of S. Then show that $2(c - 1) < a$. Finally show that this implies $a = 2c - 1$.

EXERCISE 17.27. Prove that if a and b are odd then ab is odd. Hint: Write $a = 2c + 1$ and $b = 2d + 1$ (cf. the previous exercise) and compute $(2c + 1)(2d + 1)$.

EXERCISE 17.28. Consider the following sentence: There is no bijection between $\mathbb{N}$ and $\mathbb{Z}$. Explain the mistake in the following wrong proof; this is an instance of a fallacy discussed earlier.

"*Proof.*" Assume there is a bijection $f : \mathbb{N} \to \mathbb{Z}$. Define $f(x) = x$. Then f is not surjective so it is not a bijection.

EXERCISE 17.29. Prove that there is a bijection between $\mathbb{N}$ and $\mathbb{Z}$.

CHAPTER 18

Induction

Induction is the single most important method to prove elementary theorems about the integers. (More subtle theorems, such as many of the theorems of "number theory," require more sophisticated methods.) Let $P(x)$ be a formula in the language L_{set} of sets, with free variable x. For each such $P(x)$ we have:

PROPOSITION 18.1. *(Induction Principle) Assume*
1) $P(1)$.
2) For all $n \in \mathbb{N}$ if $n \neq 1$ and $P(n-1)$ then $P(n)$.
Then for all $n \in \mathbb{N}$ we have $P(n)$.

The above is expressed, as usual, in argot. The same expressed as a sentence in L_{set} reads:

$$(P(1) \wedge (\forall x((x \in \mathbb{N}) \wedge (x \neq 0) \wedge P(x-1)) \rightarrow P(x))) \rightarrow (\forall x((x \in \mathbb{N}) \rightarrow P(x))).$$

We refer to the above as *induction on n*. For each explicit $P(x)$ this is a genuine theorem. Note that the above Proposition does not say "for all P something happens"; that would not be a sentence in the language of sets.

Proof. Let $S = \{n \in \mathbb{N}; \neg P(n)\}$. We want to show that $S = \emptyset$. Assume $S = \emptyset$ and seek a contradiction. Let m be the minimum of S. By 1) $m \neq 1$. By Exercise 17.11 $m - 1 \in \mathbb{N}$. By minimality of m, we have $P(m-1)$. By 2) we get $P(m)$, a contradiction. $\qquad\square$

EXERCISE 18.2. Define $n^2 = n \times n$ and $n^3 = n^2 \times n$ for any integer n. Prove that for any natural n there exists an integer m such that $n^3 - n = 3m$. (Later we will say that 3 divides $n^3 - n$.) Hint: Proceed by induction on n as follows. Let $P(n)$ be the sentence: for all natural n there exists an integer m such that $n^3 - n = 3m$. $P(1)$ is true because $1^3 - 1 = 3 \times 0$. Assume now that $P(n-1)$ is true i.e., $(n-1)^3 - (n-1) = 3q$ for some integer q and let us check that $P(n)$ is true i.e., that $n^3 - n = 3m$ for some integer m. The equality $(n-1)^3 - (n-1) = 3q$ reads $n^3 - 3n^2 + 3n - 1 - n + 1 = 3q$. Hence $n^3 - n = 3(n^2 - n)$ and we are done.

EXERCISE 18.3. Define $n^5 = n^3 \times n^2$. Prove that for any natural n there exists an integer m such that $n^5 - n = 5m$.

PROPOSITION 18.4. *If there exists a bijection $\{1, ..., n\} \rightarrow \{1, ..., m\}$ then $n = m$.*

Proof. We proceed by induction on n. Let $P(n)$ be the statement of the Proposition. Clearly $P(1)$ is true; cf. the Exercise below. Assume now $P(n-1)$ is true and let's prove that $P(n)$ is true. So consider a bijection $F : \{1, ..., n\} \rightarrow \{1, ..., m\}$; we want to prove that $n = m$. Let $i = F(n)$ and define the map $G : \{1, ..., n-1\} \rightarrow \{1, ..., m\}\backslash\{i\}$ by $G(j) = F(j)$ for all $1 \leq j \leq n-1$. Then clearly G is a bijection. Now consider the map $H : \{1, ..., m\}\backslash\{i\} \to \{1, ..., m - 1\}$ defined by $H(j) = j$ for $1 \leq j \leq i - 1$ and $H(j) = j - 1$ for $i + 1 \leq j \leq m$. (The definition is correct because for any $j \in \{1, ..., m\}\backslash\{i\}$ either $j \leq i - 1$ or $j \geq i + 1$; cf. Exercise 17.10.) Clearly H is a bijection. We get a bijection

$$H \circ G : \{1, ..., n - 1\} \to \{1, ..., m - 1\}.$$

Since $P(n - 1)$ is true we get $n - 1 = m - 1$. Hence $n = m$ and we are done. $\square$

EXERCISE 18.5. Check that $P(1)$ is true in the above Proposition.

REMARK 18.6. Note the general strategy of proofs by inductions. Say $P(n)$ is "about n objects." There are two steps. The first step is the verification of $P(1)$ i.e., one verifies the statement "for one object." For the second step (called the induction step) one considers a situation with n objects; one "removes" from that situation "one object" to get a "situation with $n - 1$ objects"; one uses the "induction hypothesis" $P(n - 1)$ to conclude the claim for the "situation with $n - 1$ objects." Then one tries to "go back" and prove that the claim is true for the situation with n objects. So the second step is performed by "removing" one object from an arbitrary situation with n objects and NOT by adding one object to an arbitrary situation with $n - 1$ objects. Below is an example of a fallacious reasoning by induction based on "adding" instead of "subtracting" an object.

EXAMPLE 18.7. Here is a wrong argument for the induction step in the proof of Proposition 18.4.

"Proof." Let $G : \{1, ..., n - 1\} \to \{1, ..., m - 1\}$ be any bijection and let $F : \{1, ..., n\} \to \{1, ..., m\}$ be defined by $F(i) = G(i)$ for $i \leq n - 1$ and $F(n) = m$. Clearly F is a bijection. Now by the induction hypothesis $n - 1 = m - 1$. Hence $n = m$. This ends the proof.

The mistake is that the above does not end the proof: the above argument only covers bijections $F : \{1, ..., n\} \to \{1, ..., m\}$ constructed from bijections $G : \{1, ..., n - 1\} \to \{1, ..., m - 1\}$ in the special way described above. In other words an arbitrary bijection $F : \{1, ..., n\} \to \{1, ..., m\}$ does not always arise the way we defined F in the above "proof." In some sense the mistake we just pointed out is that of defining the same constant twice (cf. Example 17.28): we were supposed to define the symbol F as being an arbitrary bijection but then we redefined F in a special way through an arbitrary G. The point is that if G is arbitrary and F is defined as above in terms of G then F will not be arbitrary (because F will always send n into m).

DEFINITION 18.8. A set A is finite if there exists an integer $n \geq 0$ and a bijection $F : \{1, ..., n\} \to A$. (Note that n is then unique by Proposition 18.4.) We write $|A| = n$ and we call this number the *cardinality* of A or the *number of elements* of A. (Note that $|\emptyset| = 0$.) If $F(i) = a_i$ we write $A = \{a_1, ..., a_n\}$. A set is infinite if it is not finite.

EXERCISE 18.9. Prove that $|\{2, 4, -6, 9, -100\}| = 5$.

EXERCISE 18.10. For any finite sets A and B we have that $A \cup B$ is finite and

$$|A \cup B| + |A \cap B| = |A| + |B|.$$

Hint: Reduce to the case $A \cap B = \emptyset$. Then if $F : \{1, ..., a\} \to A$ and $G : \{1, ..., b\} \to B$ are bijections prove that $H : \{1, ..., a + b\} \to A \cup B$ defined by $H(i) = F(i)$ for $1 \leq i \leq a$ and $H(i) = G(i - a)$ for $a + 1 \leq i \leq a + b$ is a bijection.

EXERCISE 18.11. For any finite sets A and B we have that $A \times B$ is finite and

$$|A \times B| = |A| \times |B|.$$

Hint: Induction on $|A|$.

EXERCISE 18.12. Let $F : \{1, ..., n\} \to \mathbb{Z}$ be an injective map and write $F(i) = a_i$. We refer to such a map as a (finite) family of numbers. Prove that there exists a unique map $G : \{1, ..., n\} \to \mathbb{Z}$ such that $G(1) = a_1$ and $G(k) = G(k-1) + a_k$ for $2 \leq k \leq n$. Hint: Induction on n.

DEFINITION 18.13. In the notation of the above Exercise define the (finite) sum $\sum_{i=1}^{n} a_i$ as the number $G(n)$. We also write $a_1 + ... + a_n$ for this sum. If $a_1 = ... = a_n = a$ the sum $a_1 + ... + a_n$ is written as $a + ... + a$ (n times).

EXERCISE 18.14. Prove that for any $a, b \in \mathbb{N}$ we have

$$a \times b = a + ... + a \ (b \text{ times}) = b + ... + b \ (a \text{ times}).$$

EXERCISE 18.15. Define in a similar way the (finite) product $\prod_{i=1}^{n} a_i$ (which is also denoted by $a_1 ... a_n = a_1 \times ... \times a_n$). Prove the analogues of associativity and distributivity for sums and products of families of numbers. Define a^b for $a, b \in \mathbb{N}$ and prove that $a^{b+c} = a^b \times a^c$ and $(a^b)^c = a^{bc}$.

EXERCISE 18.16. Prove that if a is an integer and n is a natural number then

$$a^n - 1 = (a - 1)(a^{n-1} + a^{n-2} + ... + a + 1).$$

Hint: Induction on n.

EXERCISE 18.17. Prove that if a is an integer and n is an integer then

$$a^{2n+1} + 1 = (a + 1)(a^{2n} - a^{2n-1} + a^{2n-2} - ... - a + 1).$$

Hint: Set $a = -b$.

EXERCISE 18.18. Prove that a subset $A \subset \mathbb{N}$ is bounded if and only if it is finite. Hint: To prove that bounded sets are finite assume this is false and let b be the minimum natural number with the property that there is a set A bounded from above by b and infinite. If $b \notin A$ then A is bounded from above by $b - 1$ (Exercise 17.10) and we are done. If $b \in A$ then, by minimality of b, there is a bijection $A \backslash \{b\} \to \{1, ..., m\}$ and one constructs a bijection $A \to \{1, ..., m + 1\}$ which is a contradiction. To prove that finite sets are bounded assume this is false and let n be minimum natural number with the property that there is a finite subset $A \subset \mathbb{N}$ of cardinality n which is not bounded. Let $F : \{1, ..., n\} \to A$ be a bijection, $a_i = F(i)$. Then $\{a_1, ..., a_{n-1}\}$ is bounded from above by some b and conclude that A is bounded from above by either b or a_n.

EXERCISE 18.19. Prove that any subset of a finite set is finite. Hint: Use the previous exercise.

DEFINITION 18.20. Let A be a set and $n \in \mathbb{N}$. Define the set A^n to be the set $A^{\{1,...,n\}}$ of all maps $\{1, ..., n\} \to A$. Call

$$A^\star = \bigcup_{n=1}^{\infty} A^n$$

the set of words with letters in A.

DEFINITION 18.21. If $f : \{1, ..., n\} \to A$ and $f(i) = a_i$ we write f as a "tuple" $(a_1, ..., a_n)$ and sometimes as a "word" $a_1...a_n$; in other words we add to the definitions of set theory the following definitions

$$f = (a_1, ..., a_n) = a_1...a_n.$$

EXERCISE 18.22. Show that the maps $A^n \times A^m \to A^{n+m}$,

$$((a_1, ..., a_n), (b_1, ..., b_m)) \mapsto (a_1, ..., a_n, b_1, ..., b_m)$$

(called concatenations), are bijections. They induce a non-injective binary operation $A^\star \times A^\star \to A^\star$, $(u, v) \to uv$. Prove that $u(vw) = (uv)w$.

CHAPTER 19

Rationals

With the integers at our disposal one can use the axioms of set theory to construct a whole array of familiar sets of numbers such as the rationals, the reals, the imaginaries, etc. We start here with the rationals.

DEFINITION 19.1. For any $a, b \in \mathbb{Z}$ with $b \neq 0$ define the fraction $\frac{a}{b}$ to be the set of all pairs (c, d) with $c, d \in \mathbb{Z}$, $d \neq 0$ such that $ad = bc$. Denote by $\mathbb{Q}$ the set of all fractions. So

$$\frac{a}{b} = \{(c, d) \in \mathbb{Z} \times \mathbb{Z}; d \neq 0, ad = bc\},$$

$$\mathbb{Q} = \{\frac{a}{b}; a, b \in \mathbb{Z}, b \neq 0\}.$$

EXAMPLE 19.2.

$$\frac{6}{10} = \{(6, 10), (-3, -5), (9, 15), ...\} \in \mathbb{Q}.$$

EXERCISE 19.3. Prove that $\frac{a}{b} = \frac{c}{d}$ if and only if $ad = bc$. Hint: Assume $ad = bc$ and let us prove that $\frac{a}{b} = \frac{c}{d}$. We need to show that $\frac{a}{b} \subset \frac{c}{d}$ and that $\frac{c}{d} \subset \frac{a}{b}$. Now if $(x, y) \in \frac{a}{b}$ then $xb = ay$; hence $xbd = ayd$. Since $ad = bc$ we get $xbd = bcy$. Hence $b(xd - cy) = 0$. Since $b \neq 0$ we have $xd - cy = 0$ hence $xd = cy$ hence $(x, y) \in \frac{c}{d}$. We proved that $\frac{a}{b} \subset \frac{c}{d}$. The other inclusion is proved similarly. So the equality $\frac{a}{b} = \frac{c}{d}$ is proved. Conversely if one assumes $\frac{a}{b} = \frac{c}{d}$ one needs to prove $ad = bc$; we leave this to the reader.

EXERCISE 19.4. On the set $A = \mathbb{Z} \times (\mathbb{Z} \backslash \{0\})$ one can consider the relation: $(a, b) \sim (c, d)$ if and only if $ad = bc$. Prove that $\sim$ is an equivalence relation. Then observe that $\frac{a}{b}$ is the equivalence class

$$\widehat{(a, b)}$$

of (a, b). Also observe that $\mathbb{Q} = A/\sim$ is the quotient of A by the relation $\sim$.

EXERCISE 19.5. Prove that the map $\mathbb{Z} \to \mathbb{Q}$, $a \mapsto \frac{a}{1}$ is injective.

DEFINITION 19.6. By abuse we identify $a \in \mathbb{Z}$ with $\frac{a}{1} \in \mathbb{Q}$ and write $\frac{a}{1} = a$; this identifies $\mathbb{Z}$ with a subset of $\mathbb{Q}$. Such identifications are very common and will be done later in similar contexts.

DEFINITION 19.7. Define $\frac{a}{b} + \frac{c}{d} = \frac{ad+bc}{bd}$, $\frac{a}{b} \times \frac{c}{d} = \frac{ac}{bd}$.

EXERCISE 19.8. Show that the above definition is correct (i.e., if $\frac{a}{b} = \frac{a'}{b'}$, $\frac{c}{d} = \frac{c'}{d'}$ then $\frac{ad+bc}{bd} = \frac{a'd'+b'c'}{b'd'}$ and similarly for the product).

EXERCISE 19.9. Prove that $\mathbb{Q}$ (with the operations $+$ and $\times$ defined above and with the elements $0, 1$) is a field.

DEFINITION 19.10. For $\frac{a}{b}, \frac{c}{d}$ with $b, d > 0$ write $\frac{a}{b} \leq \frac{c}{d}$ if $ad - bc \leq 0$. Also write $\frac{a}{b} \lneq \frac{c}{d}$ if $\frac{a}{b} \leq \frac{c}{d}$ and $\frac{a}{b} \neq \frac{c}{d}$.

EXERCISE 19.11. Prove that $\mathbb{Q}$ equipped with $\leq$ is an ordered ring but it is not a well ordered ring.

EXERCISE 19.12. Let A be a non-empty finite set and define $\mu : \mathcal{P}(A) \to \mathbb{Q}$ by

$$\mu(X) = \frac{|X|}{|A|}.$$

1) $(A, \mathcal{P}(A), \mu)$ is a finite probability measure space.
2) Prove that if $X = Y \neq A$ then X and Y are not independent.
3) Prove that if $X \cap Y = \emptyset$ and $X \neq \emptyset$, $Y \neq \emptyset$ then X and Y are not independent.
4) Prove that if $A = B \times C$, $X = B' \times C$, $Y = B \times C'$, $B' \subset B$, $C' \subset C$, then X and Y are independent.

EXERCISE 19.13. Prove by induction the following equalities:

$$1 + 2 + \ldots + n = \frac{n(n+1)}{2}.$$

$$1^2 + 2^2 + \ldots + n^2 = \frac{n(n+1)(2n+1)}{6}.$$

$$1^3 + 2^3 + \ldots + n^3 = \frac{n^2(n+1)^2}{4}.$$

CHAPTER 20

Combinatorics

Combinatorics is about counting elements in (i.e., finding cardinalities of) finite sets. The origins of combinatorics are in the work of Pascal, Jakob Bernoulli, and Leibniz; these origins are intertwined with the origins of probability theory and the early development of calculus.

DEFINITION 20.1. For $n \in \mathbb{N}$ define the factorial of n (read n factorial) by

$$n! = 1 \times 2 \times ... \times n \in \mathbb{N}.$$

Also set $0! = 1$.

DEFINITION 20.2. For $0 \leq k \leq n$ in $\mathbb{N}$ define the binomial coefficient

$$\binom{n}{k} = \frac{n!}{k!(n-k)!} \in \mathbb{Q}.$$

One also reads this "n choose k."

EXERCISE 20.3. Prove that

$$\binom{n}{k} = \binom{n}{n-k}$$

and

$$\binom{n}{0} = 1, \quad \binom{n}{1} = n.$$

EXERCISE 20.4. Prove that

$$\binom{n}{k} + \binom{n}{k+1} = \binom{n+1}{k+1}.$$

Hint: Direct computation with the definition.

EXERCISE 20.5. Prove that

$$\binom{n}{k} \in \mathbb{Z}.$$

Hint: Fix k and proceed by induction on n; use Exercise 20.4.

EXERCISE 20.6. For any a, b in any ring we have

$$(a+b)^n = \sum_{k=0}^{n} \binom{n}{k} a^k b^{n-k}.$$

Here if c is in a ring R and $m \in \mathbb{N}$ then $mc = c + ... + c$ (m times). Hint: Induction on n and use Exercise 20.4.

EXERCISE 20.7. (Subsets) Prove that if $|A| = n$ then $|\mathcal{P}(A)| = 2^n$. (A set with n elements has 2^n subsets.) Hint: Induction on n; if $A = \{a_1, ..., a_{n+1}\}$ use

$$\mathcal{P}(A) = \{B \in \mathcal{P}(A); a_{n+1} \in B\} \cup \{B \in \mathcal{P}(A); a_{n+1} \notin B\}.$$

EXERCISE 20.8. (Combinations) Let A be a set with $|A| = n$, let $0 \le k \le n$, and set

$$\text{Comb}(k, A) = \{B \in \mathcal{P}(A); |B| = k\}.$$

Prove that

$$|\text{Comb}(k, A)| = \binom{n}{k}.$$

In other words a set of n elements has exactly $\binom{n}{k}$ subsets with k elements. A subset of A having k elements is called a combination of k elements from the set A.

Hint: Fix k and proceed by induction on n. If $A = \{a_1, ..., a_{n+1}\}$ use Exercise 20.4 plus the fact that $\text{Comb}(k, A)$ can be written as

$$\{B \in \mathcal{P}(A); |B| = k, a_{n+1} \in B\} \cup \{B \in \mathcal{P}(A); |B| = k, a_{n+1} \notin B\}.$$

EXERCISE 20.9. (Permutations) For a set A let $\text{Perm}(A) \subset A^A$ be the set of all bijections $F : A \to A$. A bijection $F : A \to A$ is also called a permutation. Prove that if $|A| = n$ then

$$|\text{Perm}(A)| = n!.$$

So the exercise says that a set of n elements has $n!$ permutations. Hint: Let $|A| = |B| = n$ and let $\text{Bij}(A, B)$ be the set of all bijections $F : A \to B$; it is enough to show that $|\text{Bij}(A, B)| = n!$. Proceed by induction on n; if $A = \{a_1, ..., a_{n+1}\}$, $B = \{b_1, ..., b_{n+1}\}$ then use the fact that

$$\text{Bij}(A, B) = \bigcup_{k=1}^{n+1} \{F \in \text{Bij}(A, B); F(a_1) = b_k\}.$$

For $d \in \mathbb{N}$ and X a set let X^d be the set of all maps $\{1, ..., d\} \to X$. We identify a map $i \mapsto a_i$ with the tuple $(a_1, ..., a_d)$.

EXERCISE 20.10. (Combinations with repetition) Let

$$\text{Combrep}(n, d) = \{(x_1, ..., x_d) \in \mathbb{Z}^d; x_i \ge 0, \ x_1 + ... + x_d = n\}.$$

Prove that

$$|\text{Combrep}(n, d)| = \binom{n + d - 1}{d - 1}.$$

Hint: Let $A = \{1, ..., n + d - 1\}$. Prove that there is a bijection

$$\text{Comb}(d - 1, A) \to \text{Combrep}(n, d).$$

The bijection is given by attaching to any subset

$$\{i_1, ..., i_{d-1}\} \subset \{1, ..., n + d - 1\}$$

(where $i_1 < ... < i_{d-1}$) the tuple $(x_1, ..., x_{d-1})$ where
  1) $x_1 = |\{i \in \mathbb{Z}; 1 \le i < i_1\}|$,
  2) $x_k = |\{i \in \mathbb{Z}; i_k < i < i_{k+1}\}|$, for $2 \le k \le d - 1$, and
  3) $x_d = |\{i \in \mathbb{Z}; i_{d-1} < i \le n + d - 1\}|$.

CHAPTER 21

Sequences

DEFINITION 21.1. A sequence in a set A is a map $F : \mathbb{N} \to A$. If we write $F(n) = a_n$ we also say that $a_1, a_2, ...$ is a sequence in A or that (a_n) is a sequence in A.

THEOREM 21.2. *(Recursion theorem) Let A be a set, $a \in A$ an element, and let $F_1, F_2, ...$ be a sequence of maps $A \to A$. Then there is a unique map $G : \mathbb{N} \to A$ such that $F(1) = a$ and $G(n + 1) = F_n(G(n))$ for all $n \in \mathbb{N}$.*

Sketch of proof. We construct G as a subset of $\mathbb{N} \times A$. Let $\mathcal{Y}$ be the set of all subsets $Y \subset \mathbb{N} \times A$ with the property that

$$((1, a) \in Y) \wedge (n, x) \in Y) \to \forall n((n + 1, F_n(x)) \in Y)$$

and we define

$$G = \bigcap_{Y \in \mathcal{Y}} Y.$$

Then one proves by induction on n that for any $n \in \mathbb{N}$ there exists unique $x \in A$ such that $(n, x) \in G$. So G is a map. One then checks that G has the desired property. $\qquad \square$

Here are some applications of recursion.

PROPOSITION 21.3. *Let $(A, \leq)$ be an ordered set that has no maximal element. Then there is a sequence $F : \mathbb{N} \to A$ such that for all $n \in \mathbb{N}$ we have $F(n) < F(n + 1)$.*

Proof. Let $B = \{(a, b) \in A \times A; a < b\}$. By hypothesis the first projection $F : B \to A$, $(a, b) \mapsto a$ is surjective. By the axiom of choice there exists $G : A \to B$ such that $F \circ G = I_A$. Then $G(a) > a$ for all a. By the recursion theorem there exists $F : \mathbb{N} \to A$ such that $F(n + 1) = G(F(n))$ for all n and we are done. $\qquad \square$

EXERCISE 21.4. (Uniqueness of the ring of integers) Let Z and Z' be two well ordered rings with identities 1 and $1'$. Prove that there exists a unique ring homomorphism $F : Z \to Z'$; prove that this F is bijective and increasing. Hint: Let Z_+ be the set of all elements in Z which are > 0 and similarly for Z'. By recursion (which is, of course, valid in any well ordered ring) there is a unique $F : Z_+ \to Z'_+$ satisfying $F(1) = 1'$ and $F(n + 1) = F(n) + 1'$. Define F on Z by $F(-n) - F(n)$ for $-n \in Z_+$.

DEFINITION 21.5. A set A is countable if there exists a bijection $F : \mathbb{N} \to A$.

EXAMPLE 21.6. The set of all squares $S = \{n^2; n \in \mathbb{N}\}$ is countable; indeed $F : \mathbb{Z} \to S$, $F(n) = n^2$ is a bijection.

EXERCISE 21.7. Any subset of a countable set is countable. Hint: It is enough to show that any subset $A \subset \mathbb{N}$ is countable. Let $F \subset \mathbb{N} \times \mathbb{N}$ be the set

$$F = \{(x,y) \in \mathbb{N} \times \mathbb{N}; y = \min(A \cap \{z \in \mathbb{N}; z > x\})\}$$

which is of course a map. By the recursion theorem there exists $G : \mathbb{N} \to \mathbb{N}$ such that $G(n+1) = F(G(n))$. One checks that G is injective and its image is A.

EXERCISE 21.8. Prove that $\mathbb{N} \times \mathbb{N}$ is countable. Hint: One can find injections $\mathbb{N} \times \mathbb{N} \to \mathbb{N}$; e.g., $(m,n) \mapsto (m+n)^2 + m$.

EXERCISE 21.9. Prove that $\mathbb{Q}$ is countable.

EXAMPLE 21.10. $\mathcal{P}(\mathbb{N})$ is not countable. Indeed this is a consequence of the more general theorem we proved that there is no bijection between a set A and its power set $\mathcal{P}(A)$. However it is interesting to give a reformulation of the argument in this case (Cantor's diagonal argument). Assume $\mathcal{P}(\mathbb{N})$ is countable and seek a contradiction. Since $\mathcal{P}(\mathbb{N})$ is in bijection with $\{0,1\}^{\mathbb{N}}$ we get that there is a bijection $F : \mathbb{N} \to \{0,1\}^{\mathbb{N}}$. Denote $F(n)$ by $F_n : \mathbb{N} \to \{0,1\}$. Construct a map $G : \mathbb{N} \to \{0,1\}$ by the formula

$$G(n) = \neg(F_n(n))$$

where $\neg : \{0,1\} \to \{0,1\}$, $\neg 0 = 1$, $\neg 1 = 0$. (The definition of G does not need the recursion theorem; one can define G as a "graph" directly (check!).) Since F is surjective there exists m such that $G = F_m$. In particular:

$$G(m) = F_m(m) = \neg G(m),$$

a contradiction.

REMARK 21.11. Consider the following sentence called the continuum hypothesis:

For any set A if there exists an injection $A \to \mathcal{P}(\mathbb{N})$ then either there exists an injection $A \to \mathbb{N}$ or there exists a bijection $A \to \mathcal{P}(\mathbb{N})$.

One can ask if the above is a theorem. Answering this question (raised by Cantor) leads to important investigations in set theory. The answer (given by two theorems of Gödel and Cohen in the framework of mathematical logic rather than logic) turned out to be rather surprising; see the last part of this course.

CHAPTER 22

Reals

Real numbers have been implicitly around throughout the history of mathematics as an expression of the idea of continuity of magnitudes. What amounts to an axiomatic introduction of the reals can be found in Euclid (and is attributed to Eudoxus). The first construction of the reals from the "discrete" (i.e., from the rationals) is due to Dedekind.

DEFINITION 22.1. (Dedekind) A real number is a subset $u \subset \mathbb{Q}$ of the set $\mathbb{Q}$ of rational numbers with the following properties:
1) $u \neq \emptyset$ and $u \neq \mathbb{Q}$,
2) if $x \in u$, $y \in \mathbb{Q}$, and $x \leq y$ then $y \in u$.
Denote by $\mathbb{R}$ the set of real numbers.

EXAMPLE 22.2.
1) Any rational number $x \in \mathbb{Q}$ can be identified with the real number

$$u_x = \{y \in \mathbb{Q}; x \leq y\}.$$

It is clear that $u_x = u_{x'}$ for $x, x' \in \mathbb{Q}$ implies $x = x'$. We identify any rational number x with u_x. So we may view $\mathbb{Q} \subset \mathbb{R}$.
2) One defines, for instance, for any $n \in \mathbb{N}$, $\sqrt{n} = \{x \in \mathbb{Q}; x \geq 0, x^2 \geq n\}$.

DEFINITION 22.3. A real number $u \in \mathbb{R}$ is called irrational if $u \notin \mathbb{Q}$.

DEFINITION 22.4. If u and v are real numbers we write $u \leq v$ if and only if $v \subset u$. For $u, v \geq 0$ define

$$\begin{aligned} u + v &= \{x + y; x \in u, y \in v\} \\ u \times v = uv &= \{xy; x \in u, y \in v\}. \end{aligned}$$

Note that this extends addition and multiplication on the non-negative rationals.

EXERCISE 22.5. Naturally extend the definition of addition $+$ and multiplication $\times$ of real numbers to the case when the numbers are not necessarily ≥ 0. Prove that $(\mathbb{R}, +, \times, -, 0, 1)$ is a field. Naturally extend the order $\leq$ on $\mathbb{Q}$ to an order on $\mathbb{R}$ and prove that $\mathbb{R}$ with $\leq$ is an ordered ring.

EXERCISE 22.6. Define the sum and the product of a family of real (or complex) numbers indexed by a finite set. Hint: Use the already defined concept for integers (and hence for the rationals).

EXERCISE 22.7. Prove that $(\sqrt{n})^2 = n$.

EXERCISE 22.8. Prove that for any $r \in \mathbb{R}$ with $r > 0$ there exists a unique number $\sqrt{r} \in \mathbb{R}$ such that $\sqrt{r} > 0$ and $(\sqrt{r})^2 = r$.

EXERCISE 22.9. Prove that $\sqrt{2}$ is irrational i.e., $\sqrt{2} \notin \mathbb{Q}$. Hint: Assume there exists a rational number x such that $x^2 = 2$ and seek a contradiction. Let $a \in \mathbb{N}$ be minimal with the property that $x = \frac{a}{b}$ for some b. Now $\frac{a^2}{b^2} = 2$ hence $2b^2 = a^2$. Hence a^2 is even. Hence a is even (because if a were odd then a^2 would be odd). Hence $a = 2c$ for some integer c. Hence $2b^2 = (2c)^2 = 4c^2$. Hence $b^2 = 2c^2$. Hence b^2 is even. Hence b is even. Hence $b = 2d$ for some integer d. Hence $x = \frac{2c}{2d} = \frac{c}{d}$ and $c < a$. This contradicts the minimality of a which ends the proof.

REMARK 22.10. The above proof is probably one of the "first" proofs by contradiction in the history of mathematics; this proof appears, for instance, in Aristotle, and it is believed to have been discovered by the Pythagoreans. The irrationality of $\sqrt{2}$ was translated by the Greeks as evidence that arithmetic is insufficient to control geometry ($\sqrt{2}$ is the length of the diagonal of a square with side 1) and arguably created the first crisis in the history of mathematics, leading to a separation of algebra and geometry that lasted until Fermat and Descartes.

EXERCISE 22.11. Prove that the set

$$\{r \in \mathbb{Q}; r > 0, r^2 < 2\}$$

has no supremum in $\mathbb{Q}$.

REMARK 22.12. Later we will prove that $\mathbb{R}$ is not countable.

DEFINITION 22.13. For any $a \in \mathbb{R}$ we let $|a|$ be a or $-a$ according as $a \geq 0$ or $a \leq 0$, respectively.

EXERCISE 22.14. Prove the so-called triangle inequality:

$$|a + b| \leq |a| + |b|$$

for all $a, b \in \mathbb{R}$.

DEFINITION 22.15. For $a < b$ in $\mathbb{R}$ we define the open interval

$$(a, b) = \{c \in \mathbb{R}; a < c < b\} \subset \mathbb{R}.$$

(Not to be confused with the pair $(a, b) \in \mathbb{R} \times \mathbb{R}$ which is denoted by the same symbol.)

CHAPTER 23

Topology

Topology is about geometric properties that are invariant under *continuous* transformations. An early topological result is the formula of Descartes-Euler relating the number of vertices, edges, and faces of a convex polyhedron. (We will not discuss this here as it is surprisingly difficult to present things rigorously.) After Riemann's work on surfaces defined by algebraic functions, topology became a key feature in geometry and analysis and nowadays topological ideas are to be found everywhere in mathematics, including number theory. Here we will restrict ourselves to explaining the basic idea of continuity.

DEFINITION 23.1. A topology on a set X is a subset $\mathcal{T} \subset \mathcal{P}(X)$ of the power set of X with the following properties:
1) $\emptyset \in \mathcal{T}$ and $X \in \mathcal{T}$;
2) If $U, V \in \mathcal{T}$ then $U \cap V \in \mathcal{T}$;
3) If $(U_i)_{i \in I}$ is a family of subsets $U_i \subset X$ and if for all $i \in I$ we have $U_i \in \mathcal{T}$ then $\bigcup_{i \in I} U_i \in \mathcal{T}$.
A subset $U \subset X$ is called open if $U \in \mathcal{T}$. A subset $Z \subset X$ is called closed if $X \backslash Z$ is open. Elements of X are called points of X.

EXAMPLE 23.2. $\mathcal{T} = \mathcal{P}(X)$ is a topology on X.

EXAMPLE 23.3. $\mathcal{T} = \{\emptyset, X\} \subset \mathcal{P}(X)$ is a topology on X.

EXAMPLE 23.4. A subset $U \subset \mathbb{R}$ is called open if for any $x \in U$ there exists an open interval containing x and contained in U, $x \in (a, b) \subset U$. Let $\mathcal{T} \subset \mathcal{P}(\mathbb{R})$ be the set of all open sets of $\mathbb{R}$. Then $\mathcal{T}$ is a topology on $\mathbb{R}$; we call this the Euclidean topology.

EXERCISE 23.5. Prove that $\mathcal{T}$ in Example 23.4 is a topology.

EXERCISE 23.6. Prove that if $U, V \subset X$ then
$$\mathcal{T} = \{\emptyset, U, V, U \cup V, U \cap V, X\}$$
is a topology. Find the closed sets of X.

EXERCISE 23.7. Prove that if $(\mathcal{T}_j)_{j \in J}$ is a family of topologies $\mathcal{T}_j \subset \mathcal{P}(X)$ on X then $\bigcap_{j \in J} \mathcal{T}_j$ is a topology on X.

DEFINITION 23.8. If $\mathcal{T}_0 \subset \mathcal{P}(X)$ is a subset of the power set then the intersection
$$\mathcal{T} = \bigcap_{\mathcal{T}' \supset \mathcal{T}_0} \mathcal{T}'$$
of all topologies $\mathcal{T}'$ containing $\mathcal{T}_0$ is called the topology generated by $\mathcal{T}_0$.

EXERCISE 23.9. Let $\mathcal{T}_0 = \{U, V, W\} \subset \mathcal{P}(X)$. Find explicitly the topology generated by $\mathcal{T}_0$. Find all the closed sets in that topology.

DEFINITION 23.10. A topological space is a pair $(X, \mathcal{T})$ consisting of a set X and a topology $\mathcal{T} \subset \mathcal{P}(X)$ on X. Sometimes one writes X instead of $(X, \mathcal{T})$ if $\mathcal{T}$ is understood from context.

DEFINITION 23.11. Let X and X' be two topological spaces. A map $F : X \to X'$ is continuous if for all open $V \subset X'$ the set $F^{-1}(V) \subset X$ is open.

EXERCISE 23.12. If $\mathcal{T}$ is a topology on X and $\mathcal{T}'$ is the topology on X' defined by $\mathcal{T}' = \{\emptyset, Y\}$ then any map $F : X \to X'$ is continuous.

EXERCISE 23.13. If $\mathcal{T}$ is the topology $\mathcal{T} = \mathcal{P}(X)$ on X and $\mathcal{T}'$ is any topology on X' then any map $F : X \to X'$ is continuous.

EXERCISE 23.14. Prove that if X, X', X'' are topological spaces and $G : X \to X'$, $F : X' \to X''$ are continuous maps then the composition $F \circ G : X \to X''$ is continuous.

EXERCISE 23.15. Give an example of two topological spaces X, X' and of a bijection $F : X \to X'$ such that F is continuous but F^{-1} is not continuous. (This is to be contrasted with the situation of algebraic structures to be discussed later. See Exercise 16.20.)

Motivated by the above phenomenon, one gives the following

DEFINITION 23.16. A homeomorphism between two topological spaces is a continuous bijection whose inverse is also continuous.

DEFINITION 23.17. If X is a topological space and $Y \subset X$ is a subset then the set of all subsets of Y of the form $U \cap Y$ with U open in X form a topology on Y called the induced topology.

EXERCISE 23.18. Prove that if X is a topological space and $Y \subset X$ is open then the induced topology on Y consists of all open sets of X that are contained in Y.

DEFINITION 23.19. Let X be a topological space and let $A \subset X$ be a subset. We say that A is connected if whenever U and V are open in X with $U \cap V \cap A = \emptyset$ and $A \subset U \cup V$ it follows that $U \cap A = \emptyset$ or $V \cap A = \emptyset$.

EXERCISE 23.20. Prove that if $F : X \to X'$ is continuous and $A \subset X$ is connected then $F(A) \subset X'$ is connected.

DEFINITION 23.21. Let X be a topological space and let $A \subset X$ be a subset. A point $x \in X$ is called an accumulation point of A if for any open set U in X containing x the set $U \backslash \{x\}$ contains a point of A.

EXERCISE 23.22. Let X be a topological space and let $A \subset X$ be a subset. Prove that A is closed if and only if A contains all its accumulation points.

DEFINITION 23.23. Let X be a topological space and $A \subset X$. We say A is compact if whenever

$$A \subset \bigcup_{i \in I} U_i$$

with $(U_i)_{i \in I}$ a family of open sets in X indexed by some set I there exists a finite subset $J \subset I$ such that

$$A \subset \bigcup_{j \in J} U_j.$$

We sometimes refer to $(U_i)_{i \in I}$ as an open cover of A and to $(U_j)_{j \in J}$ as a finite open subcover. So A is compact if and only if any open cover of A has a finite open subcover.

EXERCISE 23.24. Prove that if X is a topological space and X is a finite set then it is compact.

EXERCISE 23.25. Prove that $\mathbb{R}$ is not compact in the Euclidean topology. Hint: Consider the open cover

$$\mathbb{R} = \bigcup_{n \in \mathbb{N}} (-n, n)$$

and show it has no finite open subcover.

EXERCISE 23.26. Prove that no open interval (a, b) in $\mathbb{R}$ is compact $(a < b)$.

EXERCISE 23.27. Prove that if $F : X \to X'$ is a continuous map of topological spaces and $A \subset X$ is compact then $F(A) \subset X'$ is compact.

DEFINITION 23.28. A topological space X is a Hausdorff space if for any two points $x, y \in X$ there exist open sets $U \subset X$ and $V \subset X$ such that $x \in U$, $y \in V$, and $U \cap V = \emptyset$.

EXERCISE 23.29. Prove the $\mathbb{R}$ with the Euclidean topology is a Hausdorff space.

EXERCISE 23.30. Prove that if X is a Hausdorff space, $A \subset X$, and $x \in X \backslash A$ then there exist open sets $U \subset X$ and $V \subset X$ such that $x \in U$, $A \subset V$, and $U \cap V = \emptyset$. In particular any compact subset of a Hausdorff space is closed.

Hint: For any $a \in A$ let $U_a \subset X$ and $V_a \subset X$ be open sets such that $x \in U_a$, $a \in V_a$, $U_a \cap V_a = \emptyset$. Then $(V_a)_{a \in A}$ is an open covering of A. Select $(V_b)_{b \in B}$ a finite subcover of A where $B \subset A$ is a finite set, $B = \{b_1, ..., b_n\}$. Then let

$$U = U_{b_1} \cap ... \cap U_{b_n}$$
$$V = V_{b_1} \cup ... \cup V_{b_n}.$$

DEFINITION 23.31. Let X, X' be topological spaces. Then the set $X \times X'$ may be equipped with the topology generated by the family of all sets of the form $U \times U'$ where U and U' are open in X and X' respectively. This is called the product topology on $X \times X'$. Iterating this we get a product topology on a product $X_1 \times ... \times X_n$ of n topological spaces.

EXERCISE 23.32. Prove that for any $r \in \mathbb{R}$ with $r > 0$, the set
$$D = \{(x, y) \in \mathbb{R}^2; x^2 + y^2 < r^2\}$$
is open in the product topology of $\mathbb{R}^2$.

DEFINITION 23.33. A topological manifold is a topological space X such that for any point $x \in X$ there exists an open set $U \subset X$ containing x and a homeomorphism $F : U \to V$ where $V \subset \mathbb{R}^n$ is an open set in $\mathbb{R}^n$ for the Euclidean topology. (Here U and V are viewed as topological spaces with the topologies induced from X and $\mathbb{R}^n$, respectively.)

REMARK 23.34. If $\mathcal{X}$ is a set of topological manifolds then one can consider the following relation $\sim$ on $\mathcal{X}$: for $X, X' \in \mathcal{X}$ we let $X \sim X'$ if and only if there exists a homeomorphism $X \to X'$. Then $\sim$ is an equivalence relation on $\mathcal{X}$ and one of the basic problems of topology is to "describe" the set $\mathcal{X}/\sim$ of equivalence classes in various specific cases.

More properties of the Euclidean topology of $\mathbb{R}$ will be examined in the chapter on limits.

CHAPTER 24

Imaginaries

Complex numbers (also called imaginary numbers) appeared in work of Cardano, Bombelli, d'Alembert, Gauss, and others, in relation to solving polynomial equations. The modern definition below is due to Hamilton.

DEFINITION 24.1. (Hamilton) A complex number is a pair (a, b) where $a, b \in \mathbb{R}$. We denote by $\mathbb{C}$ the set of complex numbers. Hence $\mathbb{C} = \mathbb{R} \times \mathbb{R}$. Define the sum and the product of two complex numbers by

$$
\begin{aligned}
(a, b) + (c, d) &= (a + c, b + d) \\
(a, b) \times (c, d) &= (ac - bd, ad + bc).
\end{aligned}
$$

REMARK 24.2. Identify any real number $a \in \mathbb{R}$ with the complex number $(a, 0) \in \mathbb{C}$; hence write $a = (a, 0)$. In particular $0 = (0, 0)$ and $1 = (1, 0)$.

EXERCISE 24.3. Prove that $\mathbb{C}$ equipped with $0, 1$ above and with the operations $+, \times$ above is a field. Also note that the operations $+$ and $\times$ on $\mathbb{C}$ restricted to $\mathbb{R}$ are the "old" operations $+$ and $\times$ on $\mathbb{R}$.

DEFINITION 24.4. We set $i = (0, 1)$.

REMARK 24.5. $i^2 = -1$. Indeed

$$i^2 = (0, 1) \times (0, 1) = (0 \times 0 - 1 \times 1, 0 \times 1 + 1 \times 0) = (-1, 0) = -1.$$

REMARK 24.6. For any complex number $(a, b) = a + bi$. Indeed

$$(a, b) = (a, 0) + (0, b) = (a, 0) + (b, 0)(0, 1) = a + bi.$$

DEFINITION 24.7. For any complex number $z = a + bi$ we define its absolute value

$$|z| = \sqrt{a^2 + b^2}.$$

EXERCISE 24.8. Prove the so-called triangle inequality:

$$|a + b| \le |a| + |b|$$

for all $a, b \in \mathbb{C}$.

DEFINITION 24.9. For any complex number $z = a + bi$ we define its conjugate

$$\overline{z} = a - bi.$$

(The upper bar is not to be confused with the notation used in the chapter on residues.)

EXERCISE 24.10. Prove that for any $z, w \in \mathbb{C}$ we have
1) $\overline{z + w} = \overline{z} + \overline{w}$;
2) $\overline{z \cdot w} = \overline{z} \cdot \overline{w}$;
3) $\overline{z^{-1}} = \overline{z}^{-1}$ for $z \neq 0$;
4) $z \cdot \overline{z} = |z|^2$.

DEFINITION 24.11. For any complex number $z = a + bi \in \mathbb{C}$ and any real number $r > 0$ we define the open disk with center z and radius r,

$$D(z,r) = \{w \in \mathbb{C}; |w - z| < r\} \subset \mathbb{C}.$$

A subset $U \subset \mathbb{C}$ is called open if for any $z \in U$ there exists an open disk centered at z and contained in U. Let $\mathcal{T} \subset \mathcal{P}(\mathbb{C})$ be the set of all open sets of $\mathbb{C}$.

EXERCISE 24.12. Prove that $\mathcal{T}$ is a topology on $\mathbb{C}$; we call this the Euclidean topology.

EXERCISE 24.13. Prove that $\mathbb{C}$ cannot be given the structure of an ordered ring.

CHAPTER 25

Residues

Our main aim here is to introduce some of the basic "arithmetic" of $\mathbb{Z}$. In its turn arithmetic can be used to introduce the finite rings $\mathbb{Z}/m\mathbb{Z}$ of residue classes modulo m and, in particular, the finite fields $\mathbb{F}_p = \mathbb{Z}/p\mathbb{Z}$, where p is a prime. The arithmetic of $\mathbb{Z}$ to be discussed below already appears in Euclid. Congruences and residue classes were introduced by Gauss.

DEFINITION 25.1. For integers a and b we say a divides b if there exists an integer n such that $b = an$. We write $a|b$. We also say a is a divisor of b. If a does not divide b we write $a \nmid b$.

EXAMPLE 25.2. $4|20$; $-4|20$; $6 \nmid 20$.

EXERCISE 25.3. Prove that
1) if $a|b$ and $b|c$ then $a|c$;
2) if $a|b$ and $a|c$ then $a|b+c$;
3) $a|b$ defines an order relation on $\mathbb{N}$ but not on $\mathbb{Z}$.

THEOREM 25.4. *(Euclid division)* *For any $a \in \mathbb{Z}$ and $b \in \mathbb{N}$ there exist unique $q, r \in \mathbb{Z}$ such that $a = bq + r$ and $0 \leq r < b$.*

Proof. We prove the existence of q, r. The uniqueness is left to the reader. We may assume $a \in \mathbb{N}$. We proceed by contradiction. So assume there exists b and $a \in \mathbb{N}$ such that for all $q, r \in \mathbb{Z}$ with $0 \leq r < b$ we have $a \neq qb + r$. Fix such a b. We may assume a is minimum with the above property. If $a < b$ we can write $a = 0 \times b + a$, a contradiction. If $a = b$ we can write $a = 1 \times a + 0$, a contradiction. If $a > b$ set $a' = a - b$. Since $a' < a$, there exist $q', r \in \mathbb{Z}$ such that $0 \leq r < b$ and $a' = q'b + r$. But then $a = qb + r$, where $q = q' + 1$, a contradiction. $\square$

DEFINITION 25.5. For $a \in \mathbb{Z}$ denote $\langle a \rangle$ the set $\{na; n \in \mathbb{Z}\}$ of integers divisible by a. For $a, b \in \mathbb{Z}$ denote by $\langle a, b \rangle$ the set $\{ma + nb; m, n \in \mathbb{Z}\}$ of all numbers expressible as a multiple of a plus a multiple of b.

PROPOSITION 25.6. *For any integers a, b there exists an integer c such that $\langle a, b \rangle = \langle c \rangle$.*

Proof. If $a = b = 0$ we can take $c = 0$. Assume a, b are not both 0. Then the set $S = \langle a, b \rangle \cap \mathbb{N}$ is non-empty. Let c be the minimum of S. Clearly $\langle c \rangle \subset \langle a, b \rangle$. Let us prove that $\langle a, b \rangle \subset \langle c \rangle$. Let $u = ma + nb$ and let us prove that $u \in \langle c \rangle$. By Euclidean division $u = cq + r$ with $0 \leq r < c$. We want to show $r = 0$. Assume $r \neq 0$ and seek a contradiction. Write $c = m'a + n'b$. Then $r \in \mathbb{N}$ and also

$$r = u - cq = (ma + nb) - (m'a + n'b)q = (m - m'q)a + (n - n'q)b \in \langle a, b \rangle.$$

Hence $r \in S$. But $r < c$. So c is not the minimum of S, a contradiction. $\square$

PROPOSITION 25.7. *If a and b are integers and have no common divisor > 1 then there exist integers m and n such that $ma + nb = 1$.*

Proof. By the above Proposition $\langle a, b \rangle = \langle c \rangle$ for some $c \geq 1$. In particular $c|a$ and $c|b$. The hypothesis implies $c = 1$ hence $1 \in \langle a, b \rangle$. □

One of the main definitions of number theory is

DEFINITION 25.8. *An integer p is prime if $p > 1$ and if its only positive divisors are 1 and p.*

PROPOSITION 25.9. *If p is a prime and a is an integer such that $p \nmid a$ then there exist integers m, n such that $ma + np = 1$.*

Proof. a and p have no common divisor > 1 and we conclude by Proposition 25.7. □

PROPOSITION 25.10. *(Euclid Lemma) If p is a prime and $p|ab$ for integers a and b then either $p|a$ or $p|b$.*

Proof. Assume $p|ab$, $p \nmid a$, $p \nmid b$, and seek a contradiction. By Proposition 25.9 $ma + np = 1$ for some integers m, n and $m'b + n'p = 1$ for some integers m', n'. We get

$$1 = (ma + np)(m'b + n'p) = mm'ab + p(nm' + n'm + nn').$$

Since $p|ab$ we get $p|1$, a contradiction. □

THEOREM 25.11. *(Fundamental Theorem of Arithmetic) Any integer $n > 1$ can be written uniquely as a product of primes, i.e., there exist primes $p_1, p_2, ..., p_s$, where $s \geq 1$, such that*

$$n = p_1 p_2 ... p_s.$$

Moreover any such representation is unique in the following sense: if

$$p_1 p_2 ... p_s = q_1 q_2 ... q_t$$

with p_i and q_j prime and $p_1 \leq p_2 \leq ...$, $q_1 \leq q_2 \leq ...$ then $s = t$ and $p_1 = q_1$, $p_2 = q_2$, ...

Proof. Uniqueness follows from Euclid's Lemma 25.10. To prove the existence part let S be the set of all integers > 1 which are not products of primes. We want to show $S = \emptyset$. Assume the contrary and seek a contradiction. Let n be the minimum of S. Then n is not prime. So $n = ab$ with $a, b > 1$ integers. So $a < n$ and $b < n$. So $a \notin S$ and $b \notin S$. So a and b are products of primes. So n is a product of primes, a contradiction. □

EXERCISE 25.12. Prove the uniqueness part in the above theorem.

DEFINITION 25.13. Fix an integer $m \neq 0$. Define a relation $\equiv_m$ on $\mathbb{Z}$ by $a \equiv_m b$ if and only if $m|a - b$. Say a is congruent to b mod m (or modulo m). Instead of $a \equiv_m b$ one usually writes (following Gauss):

$$a \equiv b \pmod{m}.$$

EXAMPLE 25.14. $3 \equiv 17 \pmod 7$.

EXERCISE 25.15. Prove that $\equiv_m$ is an equivalence relation. Prove that the equivalence class $\bar{a}$ of a consists of all the numbers of the form $mb + a$ where $m \in \mathbb{Z}$.

EXAMPLE 25.16. If $m = 7$ then $\bar{3} = \overline{10} = \{..., -4, 3, 10, 17, ...\}$.

DEFINITION 25.17. For the equivalence relation $\equiv_m$ on $\mathbb{Z}$ the set of equivalence classes $\mathbb{Z}/\equiv_m$ is denoted by $\mathbb{Z}/m\mathbb{Z}$. The elements of $\mathbb{Z}/m\mathbb{Z}$ are called residue classes mod m.

EXERCISE 25.18. Prove that
$$\mathbb{Z}/m\mathbb{Z} = \{\overline{0}, \overline{1}, ..., \overline{m-1}\}.$$
So the residue classes mod m are: $\overline{0}, \overline{1}, ..., \overline{m-1}$. Hint: Use Euclid division.

EXERCISE 25.19. Prove that if $a \equiv b \pmod{m}$ and $c \equiv d \pmod{m}$ then $a + c \equiv b + d \pmod{m}$ and $ac \equiv bd \pmod{m}$.

DEFINITION 25.20. Define operations $+, \times, -$ on $\mathbb{Z}/m\mathbb{Z}$ by
$$\overline{a} + \overline{b} = \overline{a+b}$$
$$\overline{a} \times \overline{b} = \overline{ab}$$
$$-\overline{a} = \overline{-1}.$$

EXERCISE 25.21. Check that the above definitions are correct, in other words that if $\overline{a} = \overline{a'}$ and $\overline{b} = \overline{b'}$ then
$$\overline{a+b} = \overline{a'+b'}$$
$$\overline{ab} = \overline{a'b'}$$
$$\overline{-a} = \overline{-a'}.$$
Furthermore check that $(\mathbb{Z}/m\mathbb{Z}, +, \times, -, \overline{0}, \overline{1})$ is a ring.

DEFINITION 25.22. If p is a prime we write $\mathbb{F}_p$ in place of $\mathbb{Z}/p\mathbb{Z}$.

EXERCISE 25.23. Prove that $\mathbb{F}_p$ is a field. Hint: Use Proposition 25.9.

p-adics

p-adic numbers were invented by Hensel. They can be viewed as an arithmetic analogue of the real numbers (because one can do analysis with these numbers in the same way in which one does analysis with real numbers). p-adic numbers can also be viewed as an arithmetic analogue of real valued functions $f(x)$ of a real variable x (because p can be viewed as an analogue of x). In any case p-adic numbers play a key role in contemporary number theory.

DEFINITION 26.1. Let p be a prime. Let $S(\mathbb{Z})$ be the set of all sequences (a_n) of integers $a_n \in \mathbb{Z}$ such that for all n

$$a_n \equiv a_{n+1} \mod p^n.$$

Say that two sequences (a_n) and (b_n) as above are equivalent (write $(a_n) \sim_p (b_n)$) if for all n

$$a_n \equiv b_n \mod p^n.$$

Denote by $[a_n]$ the equivalence class of the sequence (a_n). Denote by

$$\mathbb{Z}_p = S(\mathbb{Z})/ \sim_p$$

the set of equivalence classes. Then $\mathbb{Z}_p$ is a ring with addition and multiplication defined by

$$[a_n] + [b_n] = [a_n + b_n]$$

$$[a_n][b_n] = [a_n b_n].$$

0 and 1 in this ring are the classes of the sequences $0, 0, \ldots$ and $1, 1, \ldots$, respectively. $\mathbb{Z}_p$ is called the ring of p-adic integers. (N.B. $\mathbb{Z}_p$ is the standard notation for this object in number theory; however in many books $\mathbb{Z}_p$ is used to denote what we denoted by $\mathbb{F}_p$; one should be aware of this discrepancy between generally accepted notations.)

EXERCISE 26.2. Check that the ring axioms are satisfied by $\mathbb{Z}_p$.

EXERCISE 26.3. Consider a sequence (b_n) and the sequence

$$a_n = b_1 + p b_2 + p^2 b_3 + \ldots + p^{n-1} b_n.$$

Prove that $(a_n) \in S(\mathbb{Z})$. Also, any element in S can be written like this.

REMARK 26.4. There is a surjective natural homomorphism $\mathbb{Z}_p \to \mathbb{F}_p$, $[a_n] \mapsto \overline{a_1}$.

REMARK 26.5. We summarize the main rings of numbers we have encountered so far; they are the main types of numbers in mathematics:

$$\mathbb{Z} \quad \to \quad \mathbb{Q} \quad \to \quad \mathbb{R} \quad \to \quad \mathbb{C}$$

$$\downarrow$$

$$\mathbb{Z}_p$$

$$\downarrow$$

$$\mathbb{F}_p$$

The arrows above are the natural ring homomorphisms between these rings.

EXERCISE 26.6. Prove that there are no ring homomorphisms $\mathbb{C} \to \mathbb{F}_p$ and there are no ring homomorphisms $\mathbb{F}_p \to \mathbb{C}$. (Morally the worlds of $\mathbb{F}_p$ and $\mathbb{C}$ do not "communicate directly," although they "communicate" via $\mathbb{Z}$.)

REMARK 26.7. There exist (many) injective ring homomorphisms $\mathbb{Z}_p \to \mathbb{C}$ but they are not "natural" in any way.

One can generalize the above construction as follows:

DEFINITION 26.8. Let A be any (commutative unital) ring and let $p \in A$ denote $1 + \ldots + 1$ (p "times"). For $a, b \in A$ write $a \equiv b$ mod p^n if there exists $c \in A$ such that $a - b = p^n c$. Let $S(A)$ be the set of all sequences (a_n) of integers $a_n \in A$ such that for all n

$$a_n \equiv a_{n+1} \quad \mathrm{mod} \quad p^n.$$

Say that two sequences (a_n) and (b_n) as above are equivalent (write $(a_n) \sim_p (b_n)$) if for all n

$$a_n \equiv b_n \quad \mathrm{mod} \quad p^n.$$

Denote by $[a_n]$ the equivalence class of the sequence (a_n). Denote by

$$\widehat{A} = S(A)/\sim_p$$

the set of equivalence classes. Then $\widehat{A}$ is a ring with addition and multiplication defined by

$$[a_n] + [b_n] = [a_n + b_n]$$
$$[a_n][b_n] = [a_n b_n].$$

$\widehat{A}$ is called the p-adic completion of A. (Clearly $\widehat{\mathbb{Z}} = \mathbb{Z}_p$.)

CHAPTER 27

Groups

Our next chapters investigate a few topics in algebra. Recall that algebra is the study of algebraic structures, i.e., sets with operations on them. We already introduced, and constructed, some elementary examples of algebraic structures such as rings and, in particular, fields. With rings/fields at our disposal one can study some other fundamental algebraic objects such as groups, vector spaces, polynomials. In what follows we briefly survey some of these. We begin with groups. In some sense groups are more fundamental than rings and fields; but in order to be able to look at more interesting examples we found it convenient to postpone the discussion of groups until this point. Groups appeared in mathematics in the context of symmetries of roots of polynomial equations; cf. the work of Galois that involved finite groups. Galois' work inspired Lie who investigated differential equations in place of polynomial equations; this led to (continuous) Lie groups, in particular groups of matrices. Groups eventually penetrated most of mathematics and physics (Klein, Poincaré, Einstein, Cartan, Weyl).

DEFINITION 27.1. A group is a tuple $(G, \star, ', e)$ consisting of a set G, a binary operation $\star$ on G, a unary operation $'$ on G (write $'(x) = x'$), and an element $e \in G$ (called the identity element) such that for any $x, y, z \in G$ the following axioms are satisfied:

1) $x \star (y \star z) = (x \star y) \star z$;
2) $x \star e = e \star x = x$;
3) $x \star x' = x' \star x = e$.

If in addition $x \star y = y \star x$ for all $x, y \in G$ we say G is commutative (or Abelian in honor of Abel).

REMARK 27.2. For any group G, any element $g \in G$, and any $n \in \mathbb{Z}$ one defines $g^n \in G$ exactly as in Exercise 18.15.

EXERCISE 27.3. Check the above.

DEFINITION 27.4. Sometimes one writes $e = 1$, $x \star y = xy$, $x' = x^{-1}$, $x \star ... \star x = x^n$ ($n \geq 1$ times). In the Abelian case one sometimes writes $e = 0$, $x \star y = x + y$, $x' = -x$, $x \star ... \star x = nx$ ($n \geq 1$ times). These notations depend on the context and are justified by the following examples.

EXAMPLE 27.5. If R is a ring then R is an Abelian group with $e = 0$, $x \star y = x + y$, $x' = -x$. Hence $\mathbb{Z}, \mathbb{Z}/m\mathbb{Z}, \mathbb{F}_p, \mathbb{Z}_p, \mathbb{Q}, \mathbb{R}, \mathbb{C}$ are groups "with respect to addition."

EXAMPLE 27.6. If R is a field then $R^\times = R \backslash \{0\}$ is an Abelian group with $e = 1$, $x \star y = xy$, $x' = x^{-1}$. Hence $\mathbb{Q}^\times, \mathbb{R}^\times, \mathbb{C}^\times, \mathbb{F}_p^\times$ are groups "with respect to multiplication."

EXAMPLE 27.7. The set $\text{Perm}(X)$ of bijections $\sigma : X \to X$ from a set X into itself is a group with $e = 1_X$ (the identity map), $\sigma \star \tau = \sigma \circ \tau$ (composition), $\sigma^{-1} =$ inverse map. If $X = \{1, ..., n\}$ then one writes $S_n = \text{Perm}(X)$ and calls this group the symmetric group. If $\sigma(1) = i_1, ..., \sigma(n) = i_n$ one usually writes

$$\sigma = \begin{pmatrix} 1 & 2 & ... & n \\ i_1 & 1_2 & ... & i_n \end{pmatrix}.$$

EXERCISE 27.8. Compute

$$\begin{pmatrix} 1 & 2 & 3 & 4 & 5 \\ 3 & 2 & 5 & 4 & 3 \end{pmatrix} \circ \begin{pmatrix} 1 & 2 & 3 & 4 & 5 \\ 5 & 4 & 2 & 1 & 3 \end{pmatrix}.$$

Also compute

$$\begin{pmatrix} 1 & 2 & 3 & 4 & 5 \\ 3 & 2 & 5 & 4 & 3 \end{pmatrix}^{-1}.$$

EXAMPLE 27.9. A 2×2 matrix with coefficients in a field R is a map

$$A : \{1, 2\} \times \{1, 2\} \to R.$$

If the map is given by

$$A(1, 1) = a$$
$$A(1, 2) = b$$
$$A(2, 1) = c$$
$$A(2, 2) = d$$

we write A as

$$A = \begin{pmatrix} a & b \\ c & d \end{pmatrix}.$$

Define the sum and the product of two matrices by

$$\begin{pmatrix} a & b \\ c & d \end{pmatrix} + \begin{pmatrix} a' & b' \\ c' & d' \end{pmatrix} = \begin{pmatrix} a + a' & b + b' \\ c + c' & d + d' \end{pmatrix},$$

$$\begin{pmatrix} a & b \\ c & d \end{pmatrix} \cdot \begin{pmatrix} a' & b' \\ c' & d' \end{pmatrix} = \begin{pmatrix} aa' + bc' & ab' + bd' \\ ca' + dc' & cb' + dd' \end{pmatrix}.$$

Define the product of an element $r \in R$ with a matrix $A = \begin{pmatrix} a & b \\ c & d \end{pmatrix}$ by

$$r \cdot \begin{pmatrix} a & b \\ c & d \end{pmatrix} = \begin{pmatrix} ra & rb \\ rc & rd \end{pmatrix}.$$

For a matrix $A = \begin{pmatrix} a & b \\ c & d \end{pmatrix}$ define its determinant by

$$\det(A) = ad - bc.$$

Say that A is invertible if $\det(A) \neq 0$ and setting $\delta = \det(A)$ define the inverse of A by

$$A^{-1} = \delta^{-1} \begin{pmatrix} d & -b \\ -c & a \end{pmatrix}.$$

Define the identity matrix by

$$I = \begin{pmatrix} 1 & 0 \\ 0 & 1 \end{pmatrix}$$

and the zero matrix by

$$O = \begin{pmatrix} 0 & 0 \\ 0 & 0 \end{pmatrix}.$$

Let $M_2(R)$ be the set of all matrices and $GL_2(R)$ be the set of all invertible matrices. Then the following are true:

1) $M_2(R)$ is a group with respect to addition of matrices;

2) $GL_2(R)$ is a group with respect to multiplication of matrices; it is called the general linear group of 2×2 matrices;

3) $(A+B)C = AC + BC$ and $C(A+B) = CA + CB$ for any matrices A, B, C;

4) There exist matrices A, B such that $AB \neq BA$;

5) $\det(AB) = \det(A) \cdot \det(B)$.

EXERCISE 27.10. Prove 1), 2), 3), 4), 5) above.

EXAMPLE 27.11. Groups are examples of algebraic structures so there is a well-defined notion of homomorphism of groups (or group homomorphism). According to the general definition a group homomorphism is a map between the two groups $F : G \to G'$ such that for all $a, b \in G$:

1) $F(a \star b) = F(a) \star' F(b)$,

2) $F(a^{-1}) = F(a)^{-1}$ (this is automatic !),

3) $F(e) = e'$ (this is, again, automatic !).

Here $\star$ and $\star'$ are the operations on G and G'; similarly e and e' are the corresponding identity elements.

DEFINITION 27.12. A subset H of a group G is called a subgroup if

1) For all $a, b \in H$ we have $a \star b \in H$.

2) For all $a \in H$ we have $a^{-1} \in H$.

3) $e \in H$.

EXERCISE 27.13. Show that if H is a subgroup of G then H, with the natural operation induced from G, is a group.

EXERCISE 27.14.

1) $\mathbb{Z}$ is a subgroup of $\mathbb{Q}$.

2) $\mathbb{Q}$ is a subgroup of $\mathbb{R}$.

3) $\mathbb{R}$ is a subgroup of $\mathbb{C}$.

4) If R is a field then the set

$$SL_2(R) = \{ \begin{pmatrix} a & b \\ c & d \end{pmatrix} ; a, b \in R, \ ad - bc = 1 \}$$

is a subgroup of $GL_2(R)$; it is called the special linear group.

5) If R is a field then the set

$$SO_2(R) = \{ \begin{pmatrix} a & b \\ -b & a \end{pmatrix} ; a, b \in R, \ a^2 + b^2 = 1 \}$$

is a subgroup of $SL_2(R)$; it is called the special orthogonal group.

DEFINITION 27.15. If $F : G \to G'$ is a group homomorphism define the kernel of F,

$$Ker \ F = \{a \in G; F(a) = e'\}$$

and the image of F:

$$Im \ F = \{b \in G'; \exists a \in G, F(a) = b\}.$$

EXERCISE 27.16. Prove that $Ker\ F$ is a subgroup of G and $Im\ F$ is a subgroup of G'.

CHAPTER 28

Orders

We continue our investigation of groups and introduce the concept of order of elements in a group. (This has nothing to do with the word *order* used in the phrase *order relations*.)

DEFINITION 28.1. Let G be a group and $g \in G$; we denote by $\langle g \rangle$ the set of all elements $a \in G$ for which there exists $n \in \mathbb{Z}$ such that $a = g^n$.

EXERCISE 28.2. Prove that $\langle g \rangle$ is a subgroup of G. We call $\langle g \rangle$ the subgroup generated by g.

DEFINITION 28.3. We say that a group G is cyclic if there exists $g \in G$ such that $G = \langle g \rangle$; g is called a generator of G.

EXAMPLE 28.4. $\mathbb{Z}$ is cyclic. 1 is a generator of $\mathbb{Z}$; -1 is also a generator of $\mathbb{Z}$.

EXERCISE 28.5. Prove that $\mathbb{Q}$ is not cyclic.

DEFINITION 28.6. Let G be a group and $g \in G$. We say the order of g is infinite if $g^n \neq e$ for all $n \in \mathbb{N}$. We say the order of g is $n \in \mathbb{N}$ if:
1) $g^n = e$;
2) $g^k \neq e$ for all $k \in \mathbb{N}$ with $k < n$.
We denote by $o(g)$ the order of g.

DEFINITION 28.7. The order of a finite group G is the cardinality $|G|$ of G.

EXERCISE 28.8. The order $o(g)$ of g equals the order $|\langle g \rangle|$ of $\langle g \rangle$.

EXERCISE 28.9. g has order $n \in \mathbb{N}$ if and only if:
1') $g^n = e$;
2') If $g^N = e$ for some $N \in \mathbb{N}$ then $n|N$.
Hint: If 1') and 2') above hold then clearly g has order n. Conversely if g has order n then 1') clearly holds. To check that 2') holds use Euclidean division to write $N = nq + r$ with $0 \leq r < n$. Then $g^r = (g^n)^q g^r = g^N = e$. By condition 2) in the definition of order $r = 0$ hence $n|N$.

In what follows we say that two integers are coprime if they have no common divisor > 1.

PROPOSITION 28.10. *Assume a, b are two elements in a group such that $ab = ba$ and assume $o(a)$ and $o(b)$ are coprime. Then*

$$o(ab) = o(a)o(b).$$

Proof. Set $k = o(a)$, $l = o(b)$. Clearly, since $ab = ba$ we have

$$(ab)^{kl} = (a^k)^l (b^l)^k = e.$$

Now assume $(ab)^N = e$. Raising to power l we get $a^{Nl}b^{Nl} = e$, hence $a^{Nl} = e$, hence, by Exercise 28.9, $k|Nl$. Since k and l are coprime $k|N$ (by the Fundamental Theorem of Arithmetic). In a similar way raising $(ab)^N = e$ to power k we get $a^{Nk}b^{Nk} = e$, hence $b^{Nk} = e$, hence $l|Nk$, hence $l|N$. Again, since k and l are coprime, $l|N$ and $k|N$ imply $kl|N$ and we are done. $\square$

EXERCISE 28.11. Prove that if $o(a) = kl$ then $o(a^k) = l$.

THEOREM 28.12. *(Lagrange) If H is a subgroup of a finite group G then the order of H divides the order of G: if $n = |H|$, $m = |G|$ then $n|m$. In particular if $a \in G$ then the order $o(a)$ of a divides the order $|G|$ of the group. So if $n = |G|$ then $a^n = e$.*

Proof. For each $g \in G$ we let gH be the set of all elements of G of the form gh with $h \in H$. Let $\pi : G \to \mathcal{P}(G)$ be the map $\pi(g) = gH \in \mathcal{P}(G)$. Let $\mathcal{X} = \pi(G)$ and let $\sigma : \mathcal{X} \to G$ be any map such that $\pi \circ \sigma$ is the identity of $\mathcal{X}$. (The existence of σ follows by induction.) We claim that the map

(28.1) $\mathcal{X} \times H \to G, \quad (X, h) \mapsto \sigma(X)h, \quad X \in \mathcal{X}, \quad h \in H$

is a bijection. Assuming the claim for a moment note that the claim implies

$$|\mathcal{X}| \times |H| = |G|,$$

from which the theorem follows. Let us check the claim. To prove that 28.1 is surjective let $g \in G$. Let $g' = \sigma(gH)$. Then $g'H = \pi(g') = \pi(\sigma(gH)) = gH$. So there exists $h \in H$ such that $g'h = ge = g$; hence $g = \sigma(gH)h$ which ends the proof of surjectivity. We leave the proof of injectivity to the reader. $\square$

EXERCISE 28.13. Check the injectivity of 28.1.

THEOREM 28.14. *(Fermat's Little Theorem) For any $a \in \mathbb{Z}$ and any prime p we have*

$$a^p \equiv a \pmod{p}.$$

Proof. If $p|a$ this is clear. If $p \nmid a$ let $\overline{a}$ be the image of a in $\mathbb{F}_p^{\times}$. By Lagrange's theorem applied to the group $\mathbb{F}_p^{\times}$ we have $\overline{a}^{p-1} = \overline{1}$. Hence $a^{p-1} \equiv 1 \pmod{p}$. So $a^p \equiv a \pmod{p}$. $\square$

CHAPTER 29

Vectors

Vectors implicitly appeared in a number of contexts such as mechanics (Galileo, Newton, etc.), hypercomplex numbers (Hamilton, Cayley, etc.), algebraic number theory (Dirichlet, Kummer, Eisenstein, Kronecker, etc.), and analysis (Hilbert, Banach, etc.). They are now a basic concept in linear algebra which is itself part of abstract algebra.

DEFINITION 29.1. Let R be a field. A vector space is an Abelian group $(V, +, -, 0)$ together with a map $R \times V \to V$, $(a, v) \mapsto av$ satisfying the following conditions for all $a, b \in R$ and all $u, v \in V$:
1) $(a + b)v = av + bv$;
2) $a(u + v) = au + av$;
3) $a(bv) = (ab)v$;
4) $1v = v$.
The elements of V are called vectors.

EXAMPLE 29.2. R^n is a vector space over R viewed with the operations

$$(a_1, ..., a_n) + (b_1, ..., b_n) = (a_1 + b_1, ..., a_n + b_n),$$

$$-(a_1, ..., a_n) = (-a_1, ..., -a_n),$$

$$c(a_1, ..., a_n) = (ca_1, ..., ca_n).$$

DEFINITION 29.3. The elements $u_1, ..., u_n \in V$ are linearly independent if whenever $a_1, ..., a_n \in R$ satisfies $(a_1, ..., a_n) \neq (0, ..., 0)$ it follows that $a_1 u_1 + ... + a_n u_n \neq 0$.

DEFINITION 29.4. The elements $u_1, ..., u_n \in V$ generate V if for any $u \in V$ there exist $a_1, ..., a_n \in R$ such that $u = a_1 u_1 + ..., a_n u_n$. (We also say that u is a linear combination of $u_1, ..., u_n$.)

DEFINITION 29.5. The elements $u_1, ..., u_n \in V$ are a basis of V if they are linearly independent and generate V.

EXERCISE 29.6.
1) Show that $(-1, 1, 0)$ and $(0, 1, -1)$ are linearly independent in R^3 but they do not generate R^3.
2) Show that $(-1, 1, 0), (0, 1, -1), (1, 0, 1), (0, 2, -1)$ generate R^3 but are not linearly independent in R^3.
3) Show that $(-1, 1, 0), (0, 1, -1), (1, 0, 1)$ is a basis in R^3.

EXERCISE 29.7. If V has a basis $u_1, ..., u_n$ then the map $R^n \to V$, $(a_1, ..., a_n) \mapsto a_1 u_1 + ... + a_n u_n$ is bijective. Hint: Directly from definitions.

EXERCISE 29.8. If V is generated by $u_1, ..., u_n$ then V has a basis consisting of at most n elements. Hint: Considering a subset of $\{u_1, ..., u_n\}$ minimal with the property that it generates V we may assume that any subset obtained from $\{u_1, ..., u_n\}$

does not generate V. We claim that $u_1, ..., u_n$ are linearly independent. Assume not. Hence there exists $(a_1, ..., a_n) \neq (0, ..., 0)$ such that $a_1 u_1 + ... + a_n u_n = 0$. We may assume $a_1 = 1$. Then one checks that $u_2, ..., u_n$ generate V, contradicting minimality.

EXERCISE 29.9. Assume $R = \mathbb{F}_p$ and V has a basis with n elements. Then $|V| = p^n$.

THEOREM 29.10. *If V has a basis $u_1, ..., u_n$ and a basis $v_1, ..., v_m$ then $n = m$.*

Proof. We prove $m \leq n$; similarly one has $n \leq m$. Assume $m > n$ and seek a contradiction. Since $u_1, ..., u_n$ generate V we may write $v_1 = a_1 u_1 + ... + a_n u_n$ with not all $a_1, ..., a_n$ zero. Renumbering $u_1, ..., u_n$ we may assume $a_1 \neq 0$. Hence $v_1, u_2, ..., u_n$ generates V. Hence $v_2 = b_1 v_1 + b_2 u_2 + ... + b_n u_n$. But not all $b_2, ..., b_n$ can be zero because v_1, v_2 are linearly independent. So renumbering $u_2, ..., u_n$ we may assume $b_2 \neq 0$. So $v_1, v_2, u_3, ..., u_n$ generates V. Continuing (one needs induction) we get that $v_1, v_2, ..., v_n$ generates V. So $v_{n+1} = d_1 v_n + ... + d_n v_n$. But this contradicts the fact that $v_1, ..., v_m$ are linearly independent. $\square$

EXERCISE 29.11. Give a quick proof of the above theorem in case $R = \mathbb{F}_p$. Hint: We have $p^n = p^m$ hence $n = m$.

DEFINITION 29.12. We say V is finite dimensional (or that it has a finite basis) if there exists a basis $u_1, ..., u_n$ of V. Then we define the dimension of V to be n; write $\dim V = n$. (The definition is correct due to Theorem 29.10.)

DEFINITION 29.13. If V and W are vector spaces a map $F : V \to W$ is called linear if for all $a \in K$, $u, v \in V$ we have:
1) $F(au) = aF(u)$,
2) $F(u + v) = F(u) + F(v)$.

EXAMPLE 29.14. If $a, b, c, d, e, f \in R$ then the map $F : R^3 \to R^2$ given by
$$F(u, v, w) = (au + bv + cw, du + ev + fw)$$
is a linear map.

EXERCISE 29.15. Prove that if $F : V \to W$ is a linear map of vector spaces then $V' = F^{-1}(0)$ and $V'' = F(V)$ are vector spaces (with respect to the obvious operations). If in addition V and W are finite dimensional then V' and V'' are finite dimensional and
$$\dim V = \dim V' + \dim V''.$$
Hint: Construct corresponding bases.

EXERCISE 29.16. Give an example of a vector space that is not finite dimensional.

CHAPTER 30

Matrices

Matrices appeared in the context of linear systems of equations and were studied in the work of Leibniz, Cramer, Cayley, Eisenstein, Hamilton, Sylvester, Jordan, etc. They were later rediscovered and applied in the context of Heisenberg's matrix mechanics. Nowadays they are a standard concept in linear algebra courses.

DEFINITION 30.1. Let $m, n \in \mathbb{N}$. An $m \times n$ matrix with coefficients in a field R is a map

$$A : \{1, ..., m\} \times \{1, ..., n\} \to R.$$

If $A(i, j) = a_{ij}$ for $1 \le i \le m$, $1 \le j \le n$ then we write

$$A = (a_{ij}) = \begin{pmatrix} a_{11} & \cdots & a_{1n} \\ \cdots & \cdots & \cdots \\ a_{m1} & \cdots & a_{mn} \end{pmatrix}.$$

We denote by

$$R^{m \times n} = M_{m \times n}(R)$$

the set of all $m \times n$ matrices. We also write $M_n(R) = M_{n \times n}(R)$. Note that $R^{1 \times n}$ identifies with R^n; its elements are of the form

$$(a_1, ..., a_n)$$

and are called row matrices. Similarly the elements of $R^{m \times 1}$ are of the form

$$\begin{pmatrix} a_1 \\ \cdots \\ \cdots \\ a_m \end{pmatrix}$$

and are called column matrices. If $A = (a_{ij}) \in R^{m \times n}$ then

$$u^1 = \begin{pmatrix} a_{11} \\ \cdots \\ \cdots \\ a_{m1} \end{pmatrix}, ..., u^n = \begin{pmatrix} a_{1n} \\ \cdots \\ \cdots \\ a_{mn} \end{pmatrix}$$

are called the columns of A and we also write

$$A = (u^1, ..., u^n).$$

Similarly

$$(a_{11}, ..., a_{1n}), ..., (a_{m1}, ..., a_{mn})$$

are called the rows of A.

DEFINITION 30.2. Let $0 \in R^{m \times n}$ the matrix $0 = (z_{ij})$ with $z_{ij} = 0$ for all i, j; 0 is called the zero matrix. Let $I \in R^{n \times n}$ the matrix $I = (\delta_{ij})$ where $\delta_{ii} = 1$ for all i and $\delta_{ij} = 0$ for all $i \neq j$; I is called the identity matrix and δ_{ij} is called the Kronecker symbol.

DEFINITION 30.3. If $A = (a_{ij}), B = (b_{ij}) \in R^{m \times n}$ we define the sum

$$C = A + B \in R^{m \times n}$$

as

$$C = (c_{ij}), \quad c_{ij} = a_{ij} + b_{ij}.$$

If $A = (a_{is}) \in R^{m \times k}$, $B = (b_{sj}) \in R^{k \times n}$, we define the product

$$C = AB \in R^{m \times n}$$

as

$$C = (c_{ij}), \quad c_{ij} = \sum_{s=1}^{k} a_{is} b_{sj}.$$

EXERCISE 30.4. Prove that:
1) $R^{m \times n}$ is a group with respect to $+$.
2) $A(BC) = (AB)C$ for all $A \in R^{m \times k}$, $B \in R^{k \times l}$, $C \in R^{l \times n}$.
3) $A(B + C) = AB + AC$ for all $A \in R^{m \times k}$ and $B, C \in R^{k \times n}$.
4) $(B + C)A = BA + CA$ for all $B, C \in R^{m \times k}$ and $A \in R^{k \times n}$.
5) $AI = IA$ for all $A \in R^{n \times n}$.

EXERCISE 30.5. If $A \in R^{m \times k}$, $B \in R^{k \times n}$, and the columns of B are $b^1, ..., b^n \in R^{k \times 1}$ then the columns of AB are $Ab^1, ..., Ab^n \in R^{m \times 1}$ (where Ab^i is the product of the matrices A and b^i). In other words

$$B = (b^1, ..., b^n) \quad \Rightarrow \quad AB = (Ab^1, ..., Ab^n).$$

DEFINITION 30.6.

$$\begin{pmatrix} 1 \\ 0 \\ ... \\ 0 \end{pmatrix}, \begin{pmatrix} 0 \\ 1 \\ ... \\ 0 \end{pmatrix}, ..., \begin{pmatrix} 0 \\ 0 \\ ... \\ 1 \end{pmatrix}$$

is called the standard basis of $R^{m \times 1}$

EXERCISE 30.7. Prove that the above is indeed a basis of $R^{m \times 1}$.

Here is the link between linear maps and matrices:

DEFINITION 30.8. If $F : V \to W$ is a linear map of vector spaces and $v_1, ..., v_n$ and $w_1, ..., w_m$ are bases of V and W, respectively, then for $j = 1, ..., n$ one can write uniquely

$$F(v_j) = \sum_{i=1}^{m} a_{ij} w_i.$$

The matrix $A = (a_{ij}) \in R^{m \times n}$ is called the matrix of F with respect to the bases $v_1, ..., v_n$ and $w_1, ..., w_m$.

EXERCISE 30.9. Consider a matrix $A = (a_{ij}) \in R^{m \times n}$ and consider the map

$$F : R^{n \times 1} \to R^{m \times 1}, \quad F(u) = Au \quad \text{(product of matrices)}.$$

Then the matrix of F with respect to the canonical bases of $R^{n \times 1}$ and $R^{m \times 1}$ is A itself.

Hint: Let $e^1, ..., e^n$ be the standard basis of $R^{n \times 1}$ and let $f^1, ..., f^m$ be the standard basis of $R^{m \times 1}$. Then

$$F(e^1) = Ae^1 = \begin{pmatrix} a_{11} & \cdots & a_{1n} \\ \cdots & \cdots & \cdots \\ a_{m1} & \cdots & a_{mn} \end{pmatrix} \begin{pmatrix} 1 \\ 0 \\ \cdots \\ 0 \end{pmatrix} = \begin{pmatrix} a_{11} \\ a_{21} \\ \cdots \\ a_{m1} \end{pmatrix} = a_{11} f^1 + ... + a_{m1} f^m.$$

A similar computation can be done for $e^2, ..., e^n$.

EXERCISE 30.10. Let $F : R^{n \times 1} \to R^{m \times 1}$ be a linear map and let $A \in R^{m \times n}$ be the matrix of F with respect to the standard bases. Then for all $u \in R^{n \times 1}$ we have $F(u) = Au$.

EXERCISE 30.11. Let $G : R^{n \times 1} \to R^{k \times 1}$ and let $F : R^{k \times 1} \to R^{m \times 1}$ be linear maps. Let A be the matrix of F with respect to standard bases and let B be the matrix of G with respect to the standard bases. Then the matrix of $F \circ G$ with respect to the standard bases is AB (product of matrices). Hint: $F(G(u)) = A(Bu) = (AB)u$.

DEFINITION 30.12. If $A = (a_{ij}) \in R^{m \times n}$ is a matrix one defines the transpose of A as the matrix $A^t = (a'_{ij}) \in R^{n \times m}$ where $a'_{ij} = a_{ji}$.

EXAMPLE 30.13.

$$\begin{pmatrix} a & b & c \\ d & e & f \end{pmatrix}^t = \begin{pmatrix} a & d \\ b & e \\ c & f \end{pmatrix}.$$

EXERCISE 30.14. Prove that:
1) $(A + B)^t = A^t + B^t$;
2) $(AB)^t = B^t A^t$;
3) $I^t = I$.

CHAPTER 31

Determinants

A fundamental concept in the theory of matrices is that of determinant of a matrix. The main results are due to Cauchy, Kronecker, and Weierstrass. In spite of the computational aspect of this concept the best way to approach it is via an axiomatic method as follows.

DEFINITION 31.1. Let V and W be vector spaces over a field R and let

$$f : V^n = V \times ... \times V \to W$$

be a map. We say f is multilinear if for any $v_1, ..., v_n \in V$ and any $i \in \{1, ..., n\}$ we have:
1) If $v_i = v_i' + v_i''$ then

$$f(v_1, ..., v_n) = f(v_1, ..., v_i', ..., v_n) + f(v_1, ..., v_i'', ..., v_n).$$

2) If $v_i = cv_i'$ then

$$f(v_1, ..., v_n) = cf(v_1, ..., v_i', ..., v_n).$$

EXAMPLE 31.2. $f : R^{3\times 1} \times R^{3\times 1} \to R$ defined by

$$f\left(\begin{pmatrix} a \\ b \\ c \end{pmatrix}, \begin{pmatrix} d \\ e \\ f \end{pmatrix}\right) = ad + 3bf - ce$$

is multilinear.

DEFINITION 31.3. A multilinear map $f : V^n = V \times ... \times V \to W$ is called alternating if whenever $v_1, ..., v_n \in V$ and there exist indices $i \neq j$ such that $v_i = v_j$ we have $f(v_1, ..., v_n) = 0$.

EXAMPLE 31.4. f in Example 31.2 is not alternating. On the other hand $g : R^{2\times 1} \times R^{2\times 1} \to R$ defined by

$$f\left(\begin{pmatrix} a \\ c \end{pmatrix}, \begin{pmatrix} b \\ d \end{pmatrix}\right) = 2ad - 2bc = 2\det\begin{pmatrix} a & b \\ c & d \end{pmatrix}$$

is alternating.

LEMMA 31.5. *If $f : V^n \to W$ is multilinear alternating and $v_1, ..., v_n \in V$ then for any indices $i < j$ we have*

$$f(v_1, ..., v_i, ..., v_j, ..., v_n) = -f(v_1, ..., v_j, ..., v_i, ..., v_n).$$

Here $v_1, ..., v_j, ..., v_i, ..., v_n$ is obtained from $v_1, ..., v_i, ..., v_j, ..., v_n$ by replacing v_i with v_j and v_j with v_i while leaving all the other vs unchanged.

Proof. We have

$$f(v_1, ..., v_i + v_j, ..., v_i + v_j, ..., v_n) = f(v_1, ..., v_i, ..., v_i, ..., v_n)$$
$$+ f(v_1, ..., v_i, ..., v_j, ..., v_n)$$
$$+ f(v_1, ..., v_j, ..., v_i, ..., v_n)$$
$$+ f(v_1, ..., v_j, ..., v_j, ..., v_n).$$

Hence

$$0 = f(v_1, ..., v_i, ..., v_j, ..., v_n) + f(v_1, ..., v_j, ..., v_i, ..., v_n).$$

$\square$

EXERCISE 31.6. Let $\sigma : \{1, ..., n\} \to \{1, ..., n\}$ be a bijection. Then there exists $\epsilon(\sigma) \in \{-1, 1\}$ with the following property. Let $f : V^n \to W$ be any multilinear alternating map and $v_1, ..., v_n \in V$. Then

$$f(v_{\sigma(1)}, ..., v_{\sigma(n)}) = \epsilon(\sigma) \cdot f(v_1, ..., v_n).$$

Hint: Induction on n. For the induction step distinguish two cases: $\sigma(1) = 1$ and $\sigma(1) \neq 1$. In the first case one concludes directly by the induction hypothesis. The second case can be reduced to the first case via Lemma 31.5.

We identify $(R^{n \times 1})^n$ with $R^{n \times n}$ by identifying a tuple of columns $(b^1, ..., b^n)$ with the $n \times n$ matrix whose columns are $b^1, ..., b^n$. We denote $I = I_n$ the identity $n \times n$ matrix.

LEMMA 31.7. *There exists a multilinear alternating map*

$$f : R^{n \times n} \to R$$

such that $f(I) = 1$.

Proof. We proceed by induction on n. For n we take $f(a) = a$. Assume we constructed a multilinear alternating map

$$f_{n-1} : R^{(n-1) \times (n-1)} \to R$$

such that $f_{n-1}(I_{n-1}) = 1$. Let $A = (a_{ij})$ be an $n \times n$ matrix and let A_{ij} be the $(n-1) \times (n-1)$ matrix obtained from A by deleting the i-th row and the j-th column. Fix i and define

$$f_n(A) = \sum_{j=1}^{n} (-1)^{i+j} a_{ij} f_{n-1}(A_{ij}).$$

One easily checks that f_n is multilinear, alternating, and takes value 1 on the identity matrix I_n. $\square$

EXERCISE 31.8. Check the last sentence in the proof above.

LEMMA 31.9. *If f and g are multilinear alternating maps $R^{n \times n} \to R$ and $f(I) \neq 0$ then there exists $c \in R$ such that $g(A) = cf(A)$ for all A.*

Proof. Let $A = (a_{ij})$. Let $e^1, ..., e^n$ be the standard basis of $R^{n \times 1}$. Then

$$g(A) = g\left(\sum_{i_1} a_{i_1 1} e^{i_1}, ..., \sum_{i_n} a_{i_n n} e^{i_n}\right) = \sum_{i_1} ... \sum_{i_n} a_{i_1 1} ... a_{i_n n} g(e^{i_1}, ..., e^{i_n}).$$

The terms for which $i_1, ..., i_n$ are not distinct are zero. The terms for which $i_1, ..., i_n$ are distinct are indexed by permutations σ. By Exercise 31.6 we get

$$g(A) = \left(\sum_\sigma \epsilon(\sigma) a_{\sigma(1)1} ... a_{\sigma(n)n} \right) g(I).$$

A similar formula holds for $f(A)$ and the Lemma follows.   □

By Lemmas 31.7 and 31.9 we get:

THEOREM 31.10. *There exists a unique multilinear alternating map (called determinant)*

$$\det : R^{n \times n} \to R$$

such that $\det(I) = 1$.

EXERCISE 31.11. Using the notation in the proof of Lemma 31.7 prove that:
1) For all i we have

$$\det(A) = \sum_{j=1}^n (-1)^{i+j} a_{ij} \det(A_{ij}).$$

2) For all j we have

$$\det(A) = \sum_{i=1}^n (-1)^{i+j} a_{ij} \det(A_{ij}).$$

Hint: Use Lemma 31.9.

We also have:

THEOREM 31.12. *For any two matrices* $A, B \in R^{n \times n}$ *we have*

$$\det(AB) = \det(A) \det(B).$$

Proof. Consider the multilinear alternating map $f : R^{n \times n} \to R$ defined by

$$f(u^1, ..., u^n) = \det(Au^1, ..., Au^n)$$

for $u^1, ..., u^n \in R^{n \times 1}$. By Lemma 31.9 there exists $c \in R$ such that

$$f(u^1, ..., u^n) = c \cdot \det(u^1, ..., u^n).$$

Hence
$$\det(Au^1, ..., Au^n) = c \cdot \det(u^1, ..., u^n).$$
Setting $u^i = e^i$ we get $\det(A) = c \cdot \det(I) = c$. Setting $u^i = b^i$, the columns of B, we get $\det(AB) = c \cdot \det(B)$ and the theorem is proved.   □

EXERCISE 31.13. Prove that $\epsilon(\sigma\tau) = \epsilon(\sigma)\epsilon(\tau)$ for any permutations $\sigma, \tau \in S_n$; in other words $\epsilon : S_n \to \{1, -1\}$ is a group homomorphism.

EXERCISE 31.14. Prove that if $A \in R^{n \times n}$ is a matrix such that $\det(A) \neq 0$ then A is invertible i.e., there exists $B \in R^{n \times n}$ such that $AB = BA = I$.
Hint: Define $B = (b_{ij})$ where

$$b_{ij} = (-1)^{i+j} \det(A_{ji})$$

(notation as in Lemma 31.7). Prove that $AB = BA = I$ using Exercise 31.11.

EXERCISE 31.15. Prove that if $A \in R^{n \times n}$ is a matrix then $\det(A) = \det(A^t)$.
Hint. Use Lemma 31.9.

EXERCISE 31.16. Let R be a field.

1) Prove that the set $GL_n(R) = \{A \in R^{n \times n}; \det(A) \neq 0\}$ is a group with respect to multiplication; $GL_n(R)$ is called the general linear group.

2) Prove that the set $SL_n(R) = \{A \in R^{n \times n}; \det(A) = 1\}$ is a subgroup of $GL_n(R)$; $SL_n(R)$ is called the special linear group.

3) Prove that the set $SO_n(R) = \{A \in R^{n \times n}; \det(A) = 1, \ AA^t = I\}$ is a subgroup of $SL_n(R)$; $SO_n(R)$ is called the special orthogonal group.

Check that for $n = 2$ the above correspond to the previously defined groups $GL_2(R), SL_2(R), SO_2(R)$.

EXERCISE 31.17.

1) Prove that if a linear map $F : V \to W$ is bijective then its inverse $F^{-1} : W \to V$ is also linear. Such a map will be called an isomorphism (of vector spaces).

2) Prove that the set $GL(V)$ of all isomorphisms $V \to V$ is a group under composition.

3) Assume V has a basis $v_1, ..., v_n$ and consider the map $GL(V) \to GL_n(R)$, $F \mapsto A_F$ where A_F is the matrix of F with respect to $v_1, ..., v_n$. Prove that $GL(V) \to GL_n(R)$ is an isomorphism of groups.

CHAPTER 32

Polynomials

Determining the roots of polynomials was one of the most important motivating problems in the development of algebra, especially in the work of Cardano, Lagrange, Gauss, Abel, and Galois. Here we introduce polynomials and discuss some basic facts about their roots.

DEFINITION 32.1. Let R be a ring. We define the ring of polynomials $R[x]$ in one variable with coefficients in R as follows. An element of $R[x]$ is a map $f : \mathbb{N} \cup \{0\} \to R$, $i \mapsto a_i$ with the property that there exists $i_0 \in \mathbb{N}$ such that for all $i \geq i_0$ we have $a_i = 0$; we also write such a map as

$$f = (a_0, a_1, a_2, a_3, ...).$$

We define $0, 1 \in R[x]$ by

$$0 = (0, 0, 0, 0, ...),$$
$$1 = (1, 0, 0, 0, ...).$$

If f is as above and $g = (b_0, b_1, b_2, b_3, ...)$ then addition and multiplication are defined by

$$\begin{aligned}
f + g &= (a_0 + b_0, a_1 + b_1, a_2 + b_2, a_3 + b_3, ...), \\
fg &= (a_0 b_0, a_0 b_1 + a_1 b_0, a_0 b_2 + a_1 b_1 + a_2 b_0, a_0 b_3 + a_1 b_2 + a_2 b_1 + a_3 b_0, ...).
\end{aligned}$$

We define the degree of $f = (a_0, a_1, a_2, a_3, ...)$ as

$$deg(f) = \min\{i; a_i \neq 0\}$$

if $f \neq 0$ and $deg(0) = 0$. We also define

$$x = (0, 1, 0, 0, ...)$$

and we write

$$a = (a, 0, 0, 0, ...)$$

for any $a \in R$.

EXERCISE 32.2.
1) Prove that $R[x]$ with the operations above is a ring.
2) Prove that the map $R \to R[x]$, $a \mapsto a = (a, 0, 0, 0, ...)$ is a ring homomorphism.
3) Prove that $x^2 = (0, 0, 1, 0, ...)$, $x^3 = (0, 0, 0, 1, 0, ...)$, etc.
4) Prove that if $f = (a_0, a_1, a_2, a_3, ...)$ then

$$f = a_n x^n + ... + a_1 x + a_0$$

where $n = deg(f)$. (We also write $f = f(x)$ but we DO NOT SEE $f(x)$ as a function; this is just a notation.)

EXAMPLE 32.3. If $R = \mathbb{Z}$ then
$$(3x^2 + 5x + 1)(8x^3 + 7x^2 - 2x - 1) =$$
$$= (3 \times 8)x^5 + (3 \times 7 + 5 \times 8)x^4 + (3 \times (-2) + 5 \times 7 + 1 \times 8)x^3 + \dots.$$

DEFINITION 32.4. For any $b \in R$ we define an element $f(b) \in R$ by
$$f(b) = a_n b^n + \dots + a_1 b + a_0.$$
So for any polynomial $f \in R[x]$ we can define a map (called the polynomial map defined by the polynomial f):
$$R \to R, b \mapsto f(b).$$
(The polynomial map defined by f should not be confused with the polynomial f itself; they are two different entities.) An element $b \in R$ is called a root of f (or a zero of f) if $f(b) = 0$. (Sometimes we say "a root in R" instead of "a root.")

EXAMPLE 32.5. If $R = \mathbb{F}_2$ and we consider the polynomial $f(x) = x^2 + x \in R[x]$ then $f = (\bar{0}, \bar{1}, \bar{1}, \bar{0}, \dots) \neq (\bar{0}, \bar{0}, \bar{0}, \dots) = 0$ as an element of $R[x]$; but the polynomial map $R \to R$ defined by f sends $\bar{1} \mapsto \bar{1}^2 + \bar{1} = \bar{0}$ and $\bar{0} \mapsto \bar{0}^2 + \bar{0} = \bar{0}$ so this map is the constant map with value $\bar{0}$. This shows that different polynomials (in our case f and 0) can define the same polynomial map.

EXERCISE 32.6. Let $R = \mathbb{R}$. Show that
1) $x^2 + 1$ has no root in $\mathbb{R}$.
2) $\sqrt{\sqrt{3} + 1}$ is a root of $x^4 - 2x^2 - 1 = 0$.

REMARK 32.7. One can ask if any root in $\mathbb{C}$ of a polynomial with coefficients in $\mathbb{Q}$ can be expressed, using (possibly iterated) radicals of rational numbers. The answer to this is negative as shown by Galois in the early 19th century.

EXERCISE 32.8. Let $R = \mathbb{C}$. Show that
1) i is a root of $x^2 + 1$ in $\mathbb{C}$.
2) $\frac{1+i}{\sqrt{2}}$ is a root of $x^4 + 1 = 0$ in $\mathbb{C}$.

REMARK 32.9. Leibniz mistakenly thought that the polynomial $x^4 + 1$ should have no root in $\mathbb{C}$.

EXERCISE 32.10. Let $R = \mathbb{F}_7$. Show that:
1) $x^2 + \bar{1}$ has no root in R.
2) $\bar{2}$ is a root of $x^3 - x + \bar{1}$ in R.

EXERCISE 32.11. Let $R = \mathbb{F}_5$. Show that:
1) $\bar{2}$ is a root of $x^2 + \bar{1}$ in R.
2) $x^5 - x + 1$ has no root in $\mathbb{F}_5$.

The study of roots of polynomial functions is one of the main concerns of algebra. Here are two of the main basic theorems about roots.

THEOREM 32.12. (Lagrange) If R is a field then any polynomial of degree $d \geq 1$ has at most d roots in R.

THEOREM 32.13. (Fundamental Theorem of Algebra, Gauss) If $R = \mathbb{C}$ is the complex field then any polynomial of degree ≥ 1 has at least one root in $\mathbb{C}$.

In what follows we prove Theorem 32.12. (Theorem 32.13 is beyond the scope of this course.) We need a preparation.

DEFINITION 32.14. A polynomial $f(x) = a_n x^n + ... + a_1 x + a_0$ of degree n is monic if $a_n = 1$.

PROPOSITION 32.15. *(Long division) Let $f(x), g(x) \in R[x]$ with $g(x)$ monic of degree ≥ 1. Then there exist unique $q(x), r(x) \in R[x]$ such that*

$$f(x) = g(x)q(x) + r(x)$$

and $deg(r) < deg(g)$.

Proof. Fix g (of degree m) and let us prove by induction on n that the statement above is true if $deg(f) \leq n$. The case $deg(f) = 0$ is clear because we can then take $q(x) = 0$ and $r(x) = f(x)$. For the induction step we may take f of degree n and let $f(x) = a_n x^n + ... + a_0$, $a_n \neq 0$. We may assume $n \geq m$. Then

$$deg(f - a_n x^{n-m} g) \leq n - 1$$

so by the induction hypothesis

$$f(x) - a_n x^{n-m} g(x) = g(x)q(x) + r(x)$$

with $deg(r) < m$. So

$$f(x) = g(x)(a_n x^{n-m} + q(x)) + r(x)$$

and we are done.   □

Proof of Theorem 32.12. Assume there exists a polynomial f of degree $d \geq 1$ that has $d + 1$ roots. Choose f such that d is minimal and seek a contradiction. Let $a_1, ..., a_{d+1} \in R$ be distinct roots of f. By Long Division we can write

$$f(x) = (x - a_{d+1})g(x) + r(x)$$

with $deg(r) < deg(x - a_{d+1}) = 1$. So $deg(r) = 0$ i.e., $r(x) = c \in R$. Since $f(a_{d+1}) = 0$ we get $r(x) = 0$ hence $c = 0$. Since $0 = f(a_k) = (a_k - a_{d+1})g(a_k) + c$ for $k = 1, ..., d$ it follows that $0 = (a_k - a_{d+1})g(a_k)$. Since R is a field and $a_k - a_{d+1} \neq 0$ for $k = 1, ..., d$ it follows that $g(a_k) = 0$ for $k = 1, ..., d$. But $deg(g) = d - 1$ which contradicts the minimality of d.   □

DEFINITION 32.16. A number $\alpha \in \mathbb{C}$ is called algebraic if there exists a polynomial $f(x) = a_n x^n + ... + a_1 x + a_0$ of degree $n \geq 1$ with coefficients in $\mathbb{Q}$ such that $f(\alpha) = 0$. A number $\alpha \in \mathbb{C}$ is called transcendental if it is not algebraic.

EXAMPLE 32.17. $\sqrt{2}$ is algebraic because it is a root of $x^2 - 2$.

EXERCISE 32.18. Prove that $\sqrt{\sqrt{3} + 4} + 5\sqrt{7}$ is algebraic.

REMARK 32.19. It is not clear that transcendental numbers exist. We will check that later.

EXERCISE 32.20. Prove that the set of algebraic numbers in $\mathbb{C}$ is countable.

REMARK 32.21. The main problems about roots are:
1) Find the number of roots; in case $R = \mathbb{F}_p$ this leads to some of the most subtle problems in number theory.
2) Understand when roots of polynomials with rational coefficients, say, can be expressed by radicals; this leads to Galois theory.

DEFINITION 32.22. Let R be a ring. We defined the ring of polynomials $R[x]$ in one variable x with coefficients in R. Now $R[x]$ is again a ring so we can consider the ring of polynomials $R[x][y]$ in one variable y with coefficients in $R[x]$ which we simply denote by $R[x, y]$ and refer to as the ring of polynomials in two variables x, y with coefficients in R. Again $R[x, y]$ is a ring so we can consider the ring of polynomials $R[x, y][z]$ in one variable z with coefficients in $R[x, y]$ which we denote by $R[x, y, z]$ and which we refer to as the ring of polynomials in 3 variables x, y, z with coefficients in R, etc.

EXAMPLE 32.23.
$$3x^7y^4z - x^8 + x^4y^9z^2 + 5xyz^2 = ((x^4)y^9 + (5x)y)z^2 - ((3x^7)y^4)z - (x^8) \in \mathbb{Z}[x, y, z].$$

CHAPTER 33

Congruences

We discuss here polynomial congruences which lie at the heart of number theory. The main results below are due to Fermat, Lagrange, Euler, and Gauss.

DEFINITION 33.1. Let $f(x) \in \mathbb{Z}[x]$ be a polynomial and p a prime. An integer $c \in \mathbb{Z}$ is called a root of $f(x)$ mod p (or a solution to the congruence $f(x) \equiv 0 \ (mod \ p)$) if $f(c) \equiv 0 \ (mod \ p)$; in other words if $p|f(c)$. Let $\overline{f} \in \mathbb{F}_p[x]$ be the polynomial obtained from $f \in \mathbb{Z}[x]$ by replacing the coefficients of f with their images in $\mathbb{F}_p$. Then c is a root of f mod p if and only if the image $\overline{c}$ of c in $\mathbb{F}_p$ is a root of $\overline{f}$. We denote by $N_p(f)$ the number of roots of $f(x)$ mod p contained in $\{0, 1, ..., p-1\}$; equivalently $N_p(f)$ is the number of roots of $\overline{f}$ in $\mathbb{F}_p$. If f, g are polynomials in $\mathbb{Z}[x]$ we write $N_p(f = g)$ for $N_p(f - g)$. If $Z_p(f)$ is the set of roots of $\overline{f}$ in $\mathbb{F}_p$ then of course $N_p(f) = |Z_p(f)|$.

EXERCISE 33.2.
1) 3 is a root of $x^3 + x - 13$ mod 17.
2) Any integer a is a root of $x^p - x$ mod p; this is Fermat's Little Theorem. In particular $N_p(x^p - x) = p$, $N_p(x^{p-1} = 1) = p - 1$.
3) $N_p(ax - b) = 1$ if $p \nmid a$.
4) $N_p(x^2 = 1) = 2$ if $p \neq 2$.

PROPOSITION 33.3. *For any two polynomials $f, g \in \mathbb{Z}[x]$ we have*

$$N_p(fg) \leq N_p(f) + N_p(g).$$

Proof. Clearly $Z_p(fg) \subset Z_p(f) \cup Z_p(g)$. Hence

$$|Z_p(fg)| = |Z_p(f) \cup Z_p(g)| \leq |Z_p(f)| + |Z_p(g)|.$$

$\square$

EXERCISE 33.4. Consider the polynomials

$$f(x) = x^{p-1} - 1 \text{ and } g(x) = (x - 1)(x - 2)...(x - p + 1) \in \mathbb{Z}[x].$$

Prove that all the coefficients of the polynomial $f(x) - g(x)$ are divisible by p. Conclude that p divides the sums

$$\sum_{a=1}^{p-1} a = 1 + 2 + 3 + ... + (p - 1)$$

and

$$\sum_{1 \leq a < b \leq p-1} ab = 1 \times 2 + 1 \times 3 \times ... 1 \times (p-1) + 2 \times 3 + ... + 2 \times (p-1) + ... + (p-2) \times (p-1).$$

EXERCISE 33.5. Assume $p \geq 5$ is a prime. Prove that the numerator of any fraction that is equal to
$$1 + \frac{1}{2} + \frac{1}{3} + ... + \frac{1}{p-1}$$
is divisible by p^2.

REMARK 33.6. Fix a polynomial $f(x) \in \mathbb{Z}[x]$. Some of the deepest problems and theorems in number theory can be formulated as special cases of the following two problems:

1) Understand the set of primes p such that the congruence $f(x) \equiv 0 \ (mod \ p)$ has a solution or, equivalently, such that $p|f(c)$ for some $c \in \mathbb{Z}$.

2) Understand the set of primes p such that $p = f(c)$ for some $c \in \mathbb{Z}$.

In regards to problem 1) one would like more generally to understand the function whose value at a prime p is the number $N_p(f)$. In particular one would like to understand the set of all primes p such that $N_p(f) = k$ for a given k (equivalently such that the congruence $f(x) \equiv 0 \ (mod \ p)$ has k solutions in $\{0, 1, ..., p-1\}$). We note that if $deg(f) = 1$ the problem is trivial. For $deg(f) = 2$ the problem is already highly non-trivial although a complete answer was given by Gauss in his Quadratic Reciprocity Law (to be proved later). For the quadratic polynomial $f(x) = x^2 + 1$, for instance, we will prove below (without using quadratic reciprocity) that $p|f(c)$ for some c if and only if p is of the form $4k+1$. For $deg(f)$ arbitrary the problem (and its generalizations for polynomials $f(x, y, z, ...)$ of several variables) is essentially open and part of an array of tantalizing conjectures (part of the Langlands program) that link the function $N_p(f)$ to Fourier analysis and the theory of complex analytic functions. This is beyond the scope of our course.

In regards to problem 2), by a theorem of Dirichlet, for any linear polynomial $f(x) = ax + b$ for which a and b are coprime there exist infinitely many integers k such that $f(k)$ is prime. But it is not known, for instance, if there are infinitely many integers k such that $f(k)$ is prime when $f(x)$ is a quadratic polynomial such as $f(x) = x^2 + 1$. Problem 2) has an obvious analogue for polynomials in several variables.

The following is a direct consequence of Lagrange's Theorem 32.12:

COROLLARY 33.7. *Assume $p \equiv 1 \ (mod \ d)$. Then $N_p(x^d - 1) = d$.*

Proof. By Lagrange's Theorem $N_p(x^d - 1) \leq d$. Assume $N_p(x^d - 1) < d$ and seek a contradiction. If $p - 1 = kd$ then $x^{p-1} - 1 = (x^d - 1)g(x)$ where
$$g(x) = x^{d(k-1)} + x^{d(k-2)} + ... + x^d + 1.$$
Since by Lagrange's Theorem $N_p(g) \leq d(k-1)$ we get
$$p-1 = N_p(x^{p-1}-1) = N_p((x^d-1)g) \leq N_p(x^d-1)+N_p(g) < d+d(k-1) = dk = p-1,$$
a contradiction.   □

COROLLARY 33.8.
1) *If $p \equiv 1 \ (mod \ 4)$ then $N_p(x^2 = -1) = 2$. Equivalently any prime p of the form $4k + 1$ divides some number of the form $c^2 + 1$ where c is an integer.*
2) *If $p \equiv 3 \ (mod \ 4)$ then $N_p(x^2 = -1) = 0$. Equivalently no prime p of the form $4k + 3$ can divide a number of the form $c^2 + 1$ where c is an integer.*

Proof. 1) By Corollary 33.7 if $p \equiv 1 \ (mod \ 4)$ then $N_p(x^4 - 1) = 4$. But

$$4 = N_p(x^4 - 1) \leq N_p((x^2 - 1)(x^2 + 1)) \leq N_p(x^2 - 1) + N_p(x^2 + 1) \leq N_p(x^2 + 1) + 2$$

hence $N_p(x^2 + 1) \geq 2$ and we are done.

2) Assume $p \equiv 3 \ (mod \ 4)$ so $p = 4k + 3$ and assume $N_p(x^2 = -1) > 0$ so there exists $c \in \mathbb{Z}$ such that $c^2 \equiv -1 \ (mod \ p)$; we want to derive a contradiction. We have (by Fermat's Little Theorem) that $c^p \equiv c \ (mod \ p)$. Since $p \nmid c$ we get $c^{p-1} \equiv 1 \ (mod \ p)$. But

$$c^{p-1} \equiv c^{4k+2} \equiv (c^2)^{2k+1} \equiv (-1)^{2k+1} \equiv -1 \ (mod \ p),$$

a contradiction. $\square$

EXERCISE 33.9. Prove that:

1) If $p \equiv 1 \ (mod \ 3)$ then $N_p(x^2 + x + 1) = 2$. Equivalently any prime p of the form $3k + 1$ divides some number of the form $c^2 + c + 1$.

2) If $p \equiv 2 \ (mod \ 3)$ then $N_p(x^2 + x + 1) = 0$. Equivalently no prime p of the form $3k + 2$ can divide a number of the form $c^2 + c + 1$.

DEFINITION 33.10. Let a be an integer not divisible by a prime p. The order of a mod p is the smallest positive integer k such that $a^k \equiv 1 \ (mod \ p)$. We write $k = o_p(a)$. Clearly $o_p(a)$ equals the order $o(\bar{a})$ of the image $\bar{a}$ of a in $\mathbb{F}_p$.

DEFINITION 33.11. An integer g is a primitive root mod p if it is not divisible by p and $o_p(g) = p - 1$, equivalently, if the image $\bar{g}$ of g in $\mathbb{F}_p^\times$ is a generator of the group $\mathbb{F}_p^\times$.

EXERCISE 33.12. Prove that g is a primitive root mod p if and only if it is not divisible by p and

$$g^{(p-1)/q} \not\equiv 1 \ (mod \ p)$$

for all primes $q | p - 1$.

EXERCISE 33.13. Prove that 3 is a primitive root mod 7 but 2 is not a primitive root mod 7.

The following Theorem about the existence of primitive roots was proved by Gauss:

THEOREM 33.14. *(Gauss) If p is a prime there exists a primitive root mod p. Equivalently the group $\mathbb{F}_p^\times$ is cyclic.*

Proof. By the Fundamental Theorem of Arithmetic, $p - 1 = p_1^{e_1}...p_s^{e_s}$ with $p_1, ..., p_s$ distinct primes and $e_1, ..., e_s \geq 1$. Let $i \in \{1, ..., s\}$. By Corollary 33.7 $N_p(x^{p_i^{e_i}} - 1) = p_i^{e_i}$ and $N_p(x^{p_i^{e_i-1}} - 1) = p_i^{e_i-1}$. So $x^{p_i^{e_i}} - 1$ has a root c_i mod p which is not a root mod p of $x^{p_i^{e_i-1}} - 1$. So

$$c_i^{p_i^{e_i}} \equiv 1 \ (mod \ p),$$

$$c_i^{p_i^{e_i-1}} \not\equiv 1 \ (mod \ p).$$

It follows that the order of c_i is a divisor of $p_i^{e_i}$ but not a divisor of $p_i^{e_i-1}$. Hence

$$o_p(c_i) = p_i^{e_i}.$$

By Proposition 28.10

$$o_p(c_1...c_s) = p_1^{e_1}...p_s^{e_s} = p - 1$$

so $c_1...c_s$ is a primitive root mod p. $\square$

CHAPTER 34

Lines

We start exploring topics in geometry. Geometry is the study of shapes such as lines and planes, or, more generally, curves and surfaces, etc. There are two paths towards this study: the synthetic one and the analytic (or algebraic) one. Synthetic geometry is geometry without algebra. Analytic geometry is geometry through algebra. Synthetic geometry originates with the Greek mathematics of antiquity (e.g., the treatise of Euclid). Analytic geometry was invented by Fermat and Descartes. We already encountered the synthetic approach in the discussion of the affine plane and the projective plane which were purely combinatorial objects. Here we introduce some of the most elementary structures of analytic geometry. We start with lines. Later we will look at more complicated curves.

DEFINITION 34.1. Let R be a field. The affine plane $\mathbb{A}^2 = \mathbb{A}^2(R)$ over R is the set $R^2 = R \times R$. A point $P = (x, y)$ in the plane is an element of $R \times R$. A subset $L \subset R \times R$ is called a line if there exist $a, b, c \in R$ such that $(a, b) \neq (0, 0)$ and

$$L = \{(x, y) \in R^2; ax + by + c = 0\}.$$

We say a point P lies on the line L (or we say L passes through P) if $P \in L$. Two lines are said to be parallel if they either coincide or their intersection is empty (in the last case we say they don't meet). Three points are collinear if they lie on the same line.

DEFINITION 34.2. We sometimes write $L = L(R)$ if we want to stress that coordinates are in R.

EXERCISE 34.3. Prove that:
1) There exist 3 points which are not collinear.
2) For any two distinct points P_1 and P_2 there exists a unique line L (called sometimes $P_1 P_2$) passing through P_1 and P_2. Hint: If $P_1 = (x_1, y_1)$, $P_2 = (x_2, y_2)$, and if

$$m = (y_2 - y_1)(x_2 - x_1)^{-1}$$

then the unique line through P_1 and P_2 is:

$$L = \{(x, y) \in R \times R; y - y_1 = m(x - x_1)\}.$$

In particular any two non-parallel distinct lines meet in exactly one point.
3) Given a line L and a point P there exists exactly one line L' passing through P and parallel to L. (This is called Euclid's fifth postulate but in our exposition here this is not a postulate.)

Hence $\mathbb{A}^2 = R^2 = R \times R$ together with the set $\mathcal{L}$ of all lines (in the sense above) is an affine plane in the sense of Definition 15.41.

REMARK 34.4. Not all affine planes in the sense of Definition 15.41 are affine planes over a field in the sense above. Hilbert proved that an affine plane is the affine plane over some field if and only if the theorems of Desargues (Parts I and II) and Pappus (stated below) hold. See below for the "only if direction."

EXERCISE 34.5. Prove that any line in $\mathbb{F}_p \times \mathbb{F}_p$ has exactly p points.

EXERCISE 34.6. How many lines are there in the plane $\mathbb{F}_p \times \mathbb{F}_p$?

EXERCISE 34.7. (Desargues' Theorem, Part I) Let $A_1, A_2, A_3, A_1', A_2', A_3'$ be distinct points in the plane. Also for all $i \neq j$ assume $A_i A_j$ and $A_i' A_j'$ are not parallel and let P_{ij} be their intersection. Assume the 3 lines $A_1 A_1', A_2 A_2', A_3 A_3'$ have a point in common. Then prove that the points L_{12}, L_{13}, L_{23} are collinear (i.e., lie on some line). Hint: Consider the "space" $R \times R \times R$ and define planes and lines in this space. Prove that if two planes meet and don't coincide then they meet in a line. Then prove that through any two points in space there is a unique line and through any 3 non-collinear points there is a unique plane. Now consider the projection $R \times R \times R \to R \times R$, $(x, y, z) \mapsto (x, y)$ and show that lines project onto lines. Next show that configuration of points $A_i, A_i' \in R \times R$ can be realized as the projection of a similar configuration of points $B_i, B_i' \in R \times R \times R$ not contained in a plane. (Identifying $R \times R$ with the set of points in space with zero third coordinate we take $B_i = A_i$, $B_i' = A_i'$ for $i = 1, 2$, we let B_3 have a nonzero third coordinate, and then we choose B_3' such that the lines $B_1 B_1', B_2 B_2', B_3 B_3'$ have a point in common.) Then prove "Desargues' Theorem in Space" (by noting that if Q_{ij} is the intersection of $B_i B_j$ with $B_i' B_j'$ then Q_{ij} is in the plane containing B_1, B_2, B_3 and also in the plane containing B_1', B_2', B_3'; hence Q_{ij} is in the intersection of these planes which is a line). Finally deduce the original plane Desargues by projection.

EXERCISE 34.8. (Desargues' Theorem, Part II) Let $A_1, A_2, A_3, A_1', A_2', A_3'$ be distinct points in the plane. Assume the 3 lines $A_1 A_1', A_2 A_2', A_3 A_3'$ have a point in common or they are parallel. Assume $A_1 A_2$ is parallel to $A_1' A_2'$ and $A_1 A_3$ is parallel to $A_1' A_3'$. Prove that $A_2 A_3$ is parallel to $A_2' A_3'$. Hint: Compute coordinates. There is an alternative proof that reduces Part II to Part I by using the "projective plane over our field."

EXERCISE 34.9. (Pappus' Theorem) Let P_1, P_2, P_3 be points on a line L and let Q_1, Q_2, Q_3 be points on a line $M \neq L$. Assume the lines $P_2 Q_3$ and $P_3 Q_2$ are not parallel and let A_1 be their intersection; define A_2, A_3 similarly. Then prove that A_1, A_2, A_3 are collinear. Hint (for the case L and M meet): One can assume $L = \{(x, 0); x \in R\}$, $M = \{(0, y); y \in R\}$ (explain why). Let the points $P_i = (x_i, 0)$ and $Q_i = (0, y_i)$ and compute the coordinates of A_i. Then check that the line through A_1 and A_2 passes through A_3.

REMARK 34.10. One can identify the projective plane $(\overline{\mathbb{A}^2}, \overline{\mathcal{L}})$ attached to the affine plane $(\mathbb{A}^2, \mathcal{L})$ with the pair $(\mathbb{P}^2, \check{\mathbb{P}}^2)$ defined as follows. Let $\mathbb{P}^2 = R^3 / \sim$ where $(x, y, z) \sim (x', y', z')$ if and only if there exists $0 \neq \lambda \in R$ such that $(x', y', z') = (\lambda x, \lambda y, \lambda z)$. Denote the equivalence class of (x, y, z) by $(x : y : z)$. Identify a point (x, y) in the affine plane $\mathbb{A}^2 = R^2 = R \times R$ with the point $(x : y : 1) \in \mathbb{P}^2$. Identify a point $(x_0 : y_0 : 0)$ in the complement $\mathbb{P}^2 \backslash \mathbb{A}^2$ with the class of lines in $\mathbb{A}^2$ parallel to the line $y_0 x - x_0 y = 0$. This allows one to identify the complement $\mathbb{P}^2 \backslash \mathbb{A}^2$ with the line at infinity L_∞ of $\mathbb{A}^2$. Hence we get an identification of $\mathbb{P}^2$ with $\overline{\mathbb{A}^2}$. Finally define a line in $\mathbb{P}^2$ as a set of the form

$$\overline{L} = \{(x : y : z); ax + by + cz = 0\}.$$

So under the above identifications,

$$\overline{L} = \{(x : y : 1); ax + by + c = 0\} \cup \{(x : y : 0); ax + by = 0\} = L \cup \{\widehat{L}\}$$

where L is the line in R^2 defined by $ax + by + c = 0$. Then define $\check{\mathbb{P}}^2$ to be the set of all lines $\overline{L}$ in $\mathbb{P}^2$. We get an identification of $\check{\mathbb{P}}^2$ with $\overline{\mathcal{L}}$.

Some familiar concepts such as area and distance can be defined in the above context. Assume in what follows that R is a field such that $2 \neq 0$ and identify R^2 with $R^{2 \times 1}$.

DEFINITION 34.11. Let $P_1, P_2, P_3 \in R^{2 \times 1}$ be 3 points in the plane. Define

$$\text{area}(P_1, P_2, P_3) = \frac{1}{2} \det(P_2 - P_1, P_3 - P_1) \in R.$$

EXERCISE 34.12. Prove that
1) $\text{area}(P_{\sigma(1)} P_{\sigma(2)} P_{\sigma(3)}) = \epsilon(\sigma) \cdot \text{area}(P_1, P_2, P_3)$ for all permutations $\sigma \in S_3$.
2) $\text{area}(P_1, P_2, P_3) = 0$ if and only if P_1, P_2, P_3 are collinear.
3) Let $F : R^2 \to F^2$ be an isomorphism of vector spaces and A its matrix with respect to the canonical basis. Then $A \in SL_2(R)$ if and only if "F preserves areas" in the sense that for all $P_1, P_2, P_3 \in R^2$ we have

$$\text{area}(F(P_1), F(P_2), F(P_3)) = \text{area}(P_1, P_2, P_3).$$

4) For any $P_0 \in R^{2 \times 1}$ area is "invariant under translation by P_0" in the sense that

$$\text{area}(P_1 + P_0, P_2 + P_0, P_3 + P_0) = \text{area}(P_1, P_2, P_3).$$

DEFINITION 34.13. Let $P_1, P_2 \in R^{2 \times 1}$ be 2 points in the plane. Define the distance squared between these points as

$$\text{dist}^2(P_1, P_2) = (P_2 - P_1)^t (P_2 - P_1) \in R.$$

EXERCISE 34.14. Prove that
1) $\text{dist}^2(P_1, P_2) = \text{dist}^2(P_2, P_1)$.
2) If $R = \mathbb{R}$ then $\text{dist}^2(P_1, P_2) = 0$ if and only if $P_1 = P_2$. (Show that this may fail for other fields.)
3) Let $F : R^2 \to F^2$ be an isomorphism of vector spaces and A its matrix with respect to the canonical basis. Then $A \in SO_2(R)$ if and only if "F preserves areas" and also "preserves distances" in the sense that for all $P_1, P_2 \in R^{2 \times 1}$ we have

$$\text{dist}^2(F(P_1), F(P_2)) = \text{dist}^2(P_1, P_2).$$

4) For any $P_0 \in R^{2 \times 1}$, dist^2 is "invariant under translation by P_0" in the sense that

$$\text{dist}^2(P_1 + P_0, P_2 + P_0) = \text{dist}^2(P_1, P_2).$$

CHAPTER 35

Conics

So far we were concerned with lines in the plane. Let us discuss now "higher degree curves." We start with conics. Assume R is a field with $2 = 1 + 1 \neq 0$; equivalently R does not contain the field $\mathbb{F}_2$.

DEFINITION 35.1. The circle of center $(a, b) \in R \times R$ and radius r is the set

$$C(R) = \{(x, y) \in R \times R; (x - a)^2 + (y - b)^2 = r^2\}.$$

DEFINITION 35.2. A line is tangent to a circle if it meets it in exactly one point. (We say that the line is tangent to the circle at that point.) Two circles are tangent if they meet in exactly one point.

EXERCISE 35.3. Prove that for any circle and any point on it there is exactly one line tangent to the circle at that point.

EXERCISE 35.4. Prove that:
1) A circle and a line meet in at most 2 points.
2) Two circles meet in at most 2 points.

EXERCISE 35.5. How many points does a circle of radius 1 have if $R = \mathbb{F}_{13}$? Same problem for $\mathbb{F}_{11}$.

EXERCISE 35.6. Prove that the circle $C(R)$ with center $(0, 0)$ and radius 1 is an Abelian group with $e = (1, 0)$, $(x, y)' = (x, -y)$, and group operation

$$(x_1, y_1) \star (x_2, y_2) = (x_1 x_2 - y_1 y_2, x_1 y_2 + x_2 y_1).$$

Prove that the map

$$C(R) \to SO_2(R), \quad (a, b) \mapsto \begin{pmatrix} a & b \\ -b & a \end{pmatrix}$$

is a bijective group homomorphism. (Cf. Exercise 27.14 for $SO_2(R)$.)

EXERCISE 35.7. Consider the circle $C(\mathbb{F}_{17})$. Show that $(\bar{3}, \bar{3}), (\bar{1}, \bar{0}) \in C(\mathbb{F}_{17})$ and compute $(\bar{3}, \bar{3}) \star (\bar{1}, \bar{0})$ and $2(\bar{1}, \bar{0})$ (where the latter is, of course, $(\bar{1}, \bar{0}) \star (\bar{1}, \bar{0})$).

Circles are special cases of conics:

DEFINITION 35.8. A conic is a subset $Q \subset R \times R$ of the form

$$Q = Q(R) = \{(x, y) \in R \times R; ax^2 + bxy + cy^2 + dx + ey + f = 0\}$$

for some $(a, b, c, d, e, f) \in R \times ... \times R$, where $(a, b, c) \neq (0, 0, 0)$.

We refer to (a, b, c, d, e) as the equation of the conic and if the corresponding conic passes through a point we say that the equation of the conic passes through the point. We sometimes say "conic" instead of "equation of the conic."

EXERCISE 35.9. Prove that if 5 points are given in the plane such that no 4 of them are collinear then there exists a unique conic passing through these given 5 points. Hint: Consider the vector space of all (equations of) conics that pass through a given set S of points. Next note that if one adds a point to S the dimension of this space of conics either stays the same or drops by one. Since the space of all conics has dimension 6 it is enough to show that for $r \leq 5$ the conics passing through r points are fewer than those passing through $r - 1$ of the r points. For $r = 4$, for instance, this is done by taking a conic that is a union of 2 lines.

CHAPTER 36

Cubics

DEFINITION 36.1. Let R be a field in which $2 = 1 + 1 \neq 0$, $3 = 1 + 1 + 1 \neq 0$. Equivalently R does not contain $\mathbb{F}_2$ or $\mathbb{F}_3$. A subset $Z = Z(R) \subset R \times R$ is called an affine elliptic curve if there exist $a, b \in R$ with $4a^3 + 27b^2 \neq 0$ such that

$$Z(R) = \{(x, y) \in R \times R; y^2 = x^3 + ax + b\}.$$

We call $Z(R)$ the elliptic curve over R defined by the equation $y^2 = x^3 + ax + b$. Next we define the projective elliptic curve defined by the equation $y^2 = x^3 + ax + b$ as the set

$$E(R) = Z(R) \cup \{\infty\}$$

where ∞ is an element not belonging to $Z(R)$. (We usually drop the word "projective" and we call ∞ the point at infinity on $E(R)$.) If $(x, y) \in E(R)$ define $(x, y)' = (x, -y)$. Also define $\infty' = \infty$. Next we define a binary operation $\star$ on $E(R)$ called the chord-tangent operation; we will see that $E(R)$ becomes a group with respect to this operation. First define $(x, y) \star (x, -y) = \infty$, $\infty \star (x, y) = (x, y) \star \infty = (x, y)$, and $\infty \star \infty = \infty$. Also define $(x, 0) \star (x, 0) = \infty$. Next assume $(x_1, y_1), (x_2, y_2) \in E(R)$ with $(x_2, y_2) \neq (x_1, -y_1)$. If $(x_1, y_1) \neq (x_2, y_2)$ we let L_{12} be the unique line passing through (x_1, y_1) and (x_2, y_2). Recall that explicitly

$$L_{12} = \{(x, y) \in R \times R; y - y_1 = m(x - x_1)\}$$

where

$$m = (y_2 - y_1)(x_2 - x_1)^{-1}.$$

If $(x_1, y_1) = (x_2, y_2)$ we let L_{12} be the "line tangent to $Z(R)$ at (x_1, y_1)" which is by definition given by the same equation as before except now m is defined to be

$$m = (3x_1^2 + a)(2y_1)^{-1}.$$

(This definition is inspired by the definition of slope in analytic geometry.) Finally one defines

$$(x_1, y_1) \star (x_2, y_2) = (x_3, -y_3)$$

where (x_3, y_3) is the "third point of intersection of $E(R)$ with L_{12}"; more precisely (x_3, y_3) is defined by solving the system consisting of the equations defining $E(R)$ and L_{12} as follows: replacing y in $y^2 = x^3 + ax + b$ by $y_1 + m(x - x_1)$ we get a cubic equation in x:

$$(y_1 + m(x - x_1))^2 = x^3 + ax + b$$

which can be rewritten as

$$x^3 - m^2 x^2 + ... = 0.$$

x_1, x_2 are known to be roots of this equation. We define x_3 to be the third root which is then

$$x_3 = m^2 - x_1 - x_2;$$

so we define
$$y_3 = y_1 + m(x_3 - x_1).$$
Summarizing, the definition of (x_3, y_3) is
$$(x_3, y_3) = ((y_2 - y_1)^2 (x_2 - x_1)^{-2} - x_1 - x_2, y_1 + (y_2 - y_1)(x_2 - x_1)^{-1}(x_3 - x_1))$$
if $(x_1, y_1) \neq (x_2, y_2)$, $(x_1, y_1) \neq (x_2, -y_2)$ and
$$(x_3, y_3) = ((3x_1^2 + a)^2 (2y_1)^{-2} - x_1 - x_2, y_1 + (3x_1^2 + a)(2y_1)^{-1}(x_3 - x_1))$$
if $(x_1, y_1) = (x_2, y_2)$, $y_1 \neq 0$.
   Then $E(R)$ with the above definitions is an Abelian group.

EXERCISE 36.2. Check the last statement. (N.B. Checking associativity is a very laborious exercise.)

EXERCISE 36.3. Consider the group $E(\mathbb{F}_{13})$ defined by the equation $y^2 = x^3 + \bar{8}$. Show that $(\bar{1}, \bar{3}), (\bar{2}, \bar{4}) \in E(\mathbb{F}_{13})$ and compute $(\bar{1}, \bar{3}) \star (\bar{2}, \bar{4})$ and $2(\bar{2}, \bar{4})$ (where the latter is, of course, $(\bar{2}, \bar{4}) \star (\bar{2}, \bar{4})$).

Affine elliptic curves are special examples of cubics:

DEFINITION 36.4. A cubic is a subset $X = X(R) \subset R \times R$ of the form
$$X(R) = \{(x, y) \in R \times R; ax^3 + bx^2 y + cxy^2 + dy^3 + ex^2 + fxy + gy^2 + hx + iy + j = 0\}$$
where $(a, b, c, ..., j) \in R \times ... \times R$, $(a, b, c, d) \neq (0, ..., 0)$.

As usual we refer to the tuple $(a, b, c, ..., j)$ as the equation of a cubic (or, by abuse, simply a cubic).

EXERCISE 36.5. (Three Cubics Theorem) Prove that if two cubics meet in exactly 9 points and if a third cubic passes through 8 of the 9 points then the third cubic must pass through the 9th point. Hint: First show that if $r \leq 8$ and r points are given then the set of cubics passing through them is strictly larger than the set of cubics passing through $r - 1$ of the r points. (To show this show first that no 4 of the 9 points are on a line. Then in order to find, for instance, a cubic passing through $P_1, ..., P_7$ but not through P_8 one considers the cubics $C_i = Q_{1234i} + L_{jk}$, $\{i, j, k\} = \{5, 6, 7\}$, where Q_{1234i} is the unique conic passing through P_1, P_2, P_3, P_4, P_i and L_{jk} is the unique line through P_j and P_k. Assume C_5, C_6, C_7 all pass through P_8 and derive a contradiction as follows. Note that P_8 cannot lie on 2 of the 3 lines L_{jk} because this would force us to have 4 collinear points. So we may assume P_8 does not lie on either of the lines L_{57}, L_{67}. Hence P_8 lies on both Q_{12345} and Q_{12346}. So these conics have 5 points in common. From here one immediately gets a contradiction.) Once this is proved let $P_1, ..., P_9$ be the points of intersection of the cubics with equations F and G. We know that the space of cubics passing through $P_1, ..., P_8$ has dimension 2 and contains F and G. So any cubic in this space is a linear combination of F and G, hence will pass through P_9.

EXERCISE 36.6. (Pascal's Theorem) Let $P_1, P_2, P_3, Q_1, Q_2, Q_3$ be points on a conic C. Let A_1 be the intersection of $P_2 Q_3$ with $P_3 Q_2$, and define A_2, A_3 similarly. (Assume the lines in question are not parallel.) Then prove that A_1, A_2, A_3 are collinear. Hint: The cubics
$$Q_1 P_2 \cup Q_2 P_3 \cup Q_3 P_1 \quad \text{and} \quad P_1 Q_2 \cup P_2 Q_3 \cup P_3 Q_1$$

pass through all of the following 9 points:

$$P_1, P_2, P_3, Q_1, Q_2, Q_3, A_1, A_2, A_3.$$

On the other hand the cubic $C \cup A_2 A_3$ passes through all these points except possibly A_1. Then by the Three Cubics Theorem $C \cup A_2 A_3$ passes through A_1. Hence $A_2 A_3$ passes through A_1.

EXERCISE 36.7. Show how Pascal's Theorem implies Pappus' Theorem.

REMARK 36.8. An extended version of the Three Cubics Theorem implies the associativity of the chord-tangent operation on a cubic. (The extended version, to be used below, follows from the usual version by passing to the "projective plane"; we will not explain this proof here.) The rough idea is as follows. Let E be the elliptic curve and Q, P, R points on it. Let

$$PQ \cup E = \{P, Q, U\}$$
$$\infty U \cup E = \{\infty, U, V\}$$
$$VR \cup E = \{V, R, W\}$$
$$PR \cap E = \{P, R, X\}$$
$$\infty X \cap E = \{\infty, X, Y\}.$$

Here ∞A is the vertical passing through a point A. Note that

$$Q \star P = V, \quad V \star R = W', \quad P \star R = Y.$$

We want to show that

$$(Q \star P) \star R = Q \star (P \star R).$$

This is equivalent to

$$V \star R = Q \star Y$$

i.e., that

$$W' = Q \star Y$$

i.e., that Q, Y, W are collinear. Now the two cubics

$$E \quad \text{and} \quad PQ \cup WR \cup YX$$

both pass through the 9 points

$$P, Q, R, U, V, W, X, Y, \infty.$$

On the other hand the cubic

$$\Gamma = UV \cup PR \cup QY$$

passes through all 9 points except W. By a generalization of the Three Cubics Theorem (covering the case when one of the points is ∞) we get that Γ passes through W hence QY passes through W. The above argument only applies to chords and not to tangents. When dealing with tangents one needs to repeat the argument and look at multiplicities. So making the argument rigorous becomes technical.

CHAPTER 37

Limits

We discuss now some simple topics in analysis. Analysis is the study of "passing to the limit." The key words are sequences, convergence, limits, and later differential and integral calculus. Here we will discuss limits. Analysis emerged through work of Abel, Cauchy, Riemann, and Weierstrass, as a clarification of the early calculus of Newton, Leibniz, Euler, and Lagrange.

DEFINITION 37.1. A sequence in $\mathbb{R}$ is a map $F : \mathbb{N} \to \mathbb{R}$; if $F(n) = a_n$ we denote the sequence by $a_1, a_2, a_3, \ldots$ or by (a_n). We let $F(\mathbb{N})$ be denoted by $\{a_n; n \geq 1\}$; the latter is a subset of $\mathbb{R}$.

DEFINITION 37.2. A subsequence of a sequence $F : \mathbb{N} \to \mathbb{R}$ is a sequence of the form $F \circ G$ where $G : \mathbb{N} \to \mathbb{N}$ is an increasing map. If $a_1, a_2, a_3, \ldots$ is F then $F \circ G$ is $a_{k_1}, a_{k_2}, a_{k_3}, \ldots$ (or (a_{k_n})) where $G(n) = k_n$.

DEFINITION 37.3. A sequence (a_n) is convergent to $a_0 \in \mathbb{R}$ if for any real number $\epsilon > 0$ there exists an integer N such that for all $n \geq N$ we have $|a_n - a_0| < \epsilon$. We write $a_n \to a_0$ and we say a_0 is the limit of (a_n). A sequence is called convergent if there exists $a \in \mathbb{R}$ such that the sequence converges to a. A sequence is called divergent if it is not convergent.

EXERCISE 37.4. Prove that $a_n = \frac{1}{n}$ converges to 0.
Hint: Let $\epsilon > 0$; we need to find N such that for all $n \geq N$ we have

$$|\frac{1}{n} - 0| < \epsilon;$$

it is enough to take N to be any integer such that $N > \frac{1}{\epsilon}$.

EXERCISE 37.5. Prove that $a_n = \frac{1}{\sqrt{n}}$ converges to 0.

EXERCISE 37.6. Prove that $a_n = \frac{1}{n^2}$ converges to 0.

EXERCISE 37.7. Prove that $a_n = n$ is divergent.

EXERCISE 37.8. Prove that $a_n = (-1)^n$ is divergent.

EXERCISE 37.9. Prove that if $a_n \to a_0$ and $b_n \to b_0$ then
1) $a_n + b_n \to a_0 + b_0$
2) $a_n b_n \to a_0 b_0$.
If in addition $b_0 \neq 0$ then there exists N such that for all $n \geq N$ we have $b_n \neq 0$; moreover if $b_n \neq 0$ for all n then
3) $\frac{a_n}{b_n} \to \frac{a_0}{b_0}$.
Hint for 1: Consider any $\epsilon > 0$. Since $a_n \to a_0$ there exists N_a such that for all $n \geq N_a$ we have $|a_n - a_0| < \frac{\epsilon}{2}$. Since $b_n \to b_0$ there exists N_b such that for all

$n \geq N_b$ we have $|b_n - b_0| < \frac{\epsilon}{2}$. Let $N = \max\{N_a, N_b\}$ be the maximum between N_a and N_b. Then for all $n \geq N$ we have

$$|(a_n + b_n) - (a_0 + b_0)| \leq |a_n - a_0| + |b_n - b_0| < \frac{\epsilon}{2} + \frac{\epsilon}{2} = \epsilon.$$

EXERCISE 37.10. Prove that if $a_n \to a$, $b_n \to b$, and $a_n \leq b_n$ for all $n \geq 1$ then $a \leq b$.

DEFINITION 37.11. A sequence F is bounded if the set $F(\mathbb{N}) \subset \mathbb{R}$ is bounded.

DEFINITION 37.12. A sequence F is increasing if F is increasing. A sequence F is decreasing if $-F$ is increasing.

DEFINITION 37.13. A sequence (a_n) is Cauchy if for any real $\epsilon > 0$ there exists an integer N such that for all integers $m, n \geq N$ we have $|a_n - a_m| < \epsilon$.

EXERCISE 37.14. Prove that any convergent sequence is Cauchy.

EXERCISE 37.15. Prove the following statements in the prescribed order:
1) Any Cauchy sequence is bounded.
2) Any bounded sequence contains a sequence which is either increasing or decreasing.
3) Any bounded sequence which is either increasing or decreasing is convergent.
4) Any Cauchy sequence which contains a convergent subsequence is itself convergent.
5) Any Cauchy sequence is convergent.

Hints: For 1 let $\epsilon = 1$, let N correspond to this ϵ, and get that $|a_n - a_N| < 1$ for all $n \geq N$; conclude from here. For 2 consider the sets $A_n = \{a_m; m \geq n\}$. If at least one of these sets has no maximal element we get an increasing subsequence by Proposition 21.3. If each A_n has a maximal element b_n then $b_n = a_{k_n}$ for some k_n and the subsequence a_{k_n} is decreasing. For 3 we view each $a_n \in \mathbb{R}$ as a Dedekind cut i.e., as a subset $a_n \subset \mathbb{Q}$; the limit will be either the union of the sets a_n or the intersection. Statement 4 is easy. Statement 5 follows by combining the previous statements.

EXERCISE 37.16. Prove that any subset in $\mathbb{R}$ which is bounded from below has an infimum; and any subset in $\mathbb{R}$ which is bounded from above has a supremum.

DEFINITION 37.17. A function $F : \mathbb{R} \to \mathbb{R}$ is continuous at a point $a_0 \in \mathbb{R}$ if for any sequence (a_n) converging to a_0 we have that the sequence $(F(a_n))$ converges to $F(a_0)$.

EXERCISE 37.18. (ϵ and δ criterion). Prove that a function $F : \mathbb{R} \to \mathbb{R}$ is continuous at a_0 if and only if for any real $\epsilon > 0$ there exists a real $\delta > 0$ such that for any $a \in \mathbb{R}$ with $|a - a_0| < \delta$ we have $|F(a) - F(a_0)| < \epsilon$.

EXERCISE 37.19. Prove that a function $F : \mathbb{R} \to \mathbb{R}$ is continuous (for the Euclidean topology on both the source and the target) if and only if it is continuous at any point of $\mathbb{R}$.

EXERCISE 37.20. Prove that any polynomial function $f : \mathbb{R} \to \mathbb{R}$ (i.e., any function of the form $a \mapsto f(a)$ where f is a polynomial) is continuous.

EXERCISE 37.21. Prove that $\mathbb{R}$ with the Euclidean topology is connected.

Hint: Assume $\mathbb{R} = A \cup B$ with A, B open, non-empty, and disjoint, and seek a contradiction. Let $a \in A$ and $b \in B$. Assume $a \leq b$; the case $b \leq a$ is similar.

Show that there exists sequences (a_n) and (b_n), the first increasing, the second decreasing, with $a_n \leq b_n$ and $b_n - a_n \to 0$. (To check this use recursion to define a_{n+1}, b_{n+1} in terms of a_n, b_n by the following rule: if $c_n = \frac{a_n + b_n}{2}$ then set $a_{n+1} = c_n$ and $b_{n+1} = b_n$ in case $c_n \in A$; and set $a_{n+1} = a_n$ and $b_{n+1} = c_n$ in case $c_n \in B$.) Note that $a_n \to a_0$ and $b_n \to b_0$ and $a_0 = b_0$. Since A, B are open and disjoint they are closed. So $a_0 \in A$ and $b_0 \in B$. But this contradicts the fact that A and B are disjoint.

DEFINITION 37.22. For $a, b \in \mathbb{R}$ the closed interval $[a, b] \subset \mathbb{R}$ is defined as

$$[a, b] = \{x \in \mathbb{R}; a \leq x \leq b\}.$$

EXERCISE 37.23. Prove that $[a, b]$ are closed in the Euclidean topology.

EXERCISE 37.24. Prove that the open intervals (a, b) and the closed intervals $[a, b]$ are connected in $\mathbb{R}$.

EXERCISE 37.25. (Heine-Borel Theorem) Prove that any closed interval in $\mathbb{R}$ is compact. Hint: Assume $[a, b]$ is not compact and derive a contradiction as follows. We know $[a, b]$ has an open covering $(U_i)_{i \in I}$ that does not have a finite open subcovering. Show that there exists sequences (a_n) and (b_n), the first increasing, the second decreasing, with $a_n \leq b_n$ and $b_n - a_n \to 0$, such that $[a_n, b_n]$ cannot be covered by finitely many U_is. (To check this use recursion to define a_{n+1}, b_{n+1} in terms of a_n, b_n by the following rule: let $c_n = \frac{a_n + b_n}{2}$; then at least one of the two intervals $[a_n, c_n]$ or $[c_n, b_n]$ cannot be covered by finitely many U_is; if this is the case with the first interval then set $a_{n+1} = a_n$ and $b_{n+1} = c_n$; in the other case set $a_{n+1} = c_n$ and $b_{n+1} = b_n$.) Note that $a_n \to a_0$ and $b_n \to b_0$ and $a_0 = b_0$. But $a_0 = b_0$ is in one of the U_is; this U_i will completely contain one of the intervals $[a_n, b_n]$ which is a contradiction.

CHAPTER 38

Series

DEFINITION 38.1. Let (a_n) be a sequence and $s_n = \sum_{k=1}^{n} a_k$. The sequence (s_n) is called the sequence of partial sums. If (s_n) is convergent to some s we say $\sum_{k=1}^{\infty} a_n$ is a convergent series and that this series converges to s; we write

$$\sum_{k=1}^{\infty} a_n = s.$$

If the sequence (s_n) is divergent we say that $\sum_{k=1}^{\infty} a_n$ is a divergent series.

EXERCISE 38.2. Prove that

$$\sum_{n=1}^{\infty} \frac{1}{n(n+1)} = 1.$$

Hint: Start with the equality

$$\frac{1}{n(n+1)} = \frac{1}{n} - \frac{1}{n+1}$$

and compute

$$\sum_{n=1}^{N} \frac{1}{n(n+1)} = 1 - \frac{1}{N}.$$

EXERCISE 38.3. Prove that the series

$$\sum_{n=1}^{\infty} \frac{1}{n^2}$$

is convergent.

Hint: Prove the sequence of partial sums is bounded using the inequality

$$\frac{1}{n^2} \le \frac{1}{n(n+1)}$$

plus Exercise 38.2.

EXERCISE 38.4. Prove that the series

$$\sum_{n=1}^{\infty} \frac{1}{n^k}$$

is convergent for $k \ge 3$.

EXERCISE 38.5. Prove that the series

$$\sum_{n=1}^{\infty} \frac{1}{n}$$

is divergent. This series is called the harmonic series.

Hint: Assume the series is convergent. Then the sequence of partial sums is convergent hence Cauchy. Get a contradiction from the inequality:

$$\frac{1}{2^n+1}+\frac{1}{2^n+2}+\frac{1}{2^n+3}+\dots+\frac{1}{2^n+2^n}>2^n\times\frac{1}{2^n+2^n}=\frac{1}{2}.$$

EXERCISE 38.6. Prove that $a^n\to 0$ if $|a|<1$.

Hint: We may assume $0<a<1$. Note that (a^n) is decreasing. Since it is bounded it is convergent. Let α be its limit. Assume $\alpha\neq 0$ and get a contradiction by noting that

$$\frac{1}{a}=\frac{a^n}{a^{n+1}}\to\frac{\alpha}{\alpha}=1.$$

EXERCISE 38.7. Prove that

$$\sum_{n=1}^{\infty}a^n=\frac{1}{1-a}$$

if $|a|<1$.

EXERCISE 38.8. Prove that the series

$$\sum_{n=0}^{\infty}\frac{a^n}{n!}$$

is convergent for all $a\in\mathbb{R}$; its limit is denoted by $e^a=\exp(a)$; $e=e^1$ is called the Euler number; the map

$$\mathbb{R}\to\mathbb{R},\quad a\mapsto\exp(a)$$

is called the exponential map. Prove that

$$\exp(a+b)=\exp(a)\exp(b).$$

EXERCISE 38.9. Prove that the function $\exp:\mathbb{R}\to\mathbb{R}$ is continuous.

Hint: Let $a\in\mathbb{R}$ and $\epsilon>0$. It is enough to show that there exists $\delta>0$ such that if $|b-a|<\delta$ then $|\exp(b)-\exp(a)|<\epsilon$. Show that there is a δ such that for any b with $|b-a|<\delta$ there exists an n such that for all $m\geq n$

$$|\sum_{k=0}^{m}\frac{a^k}{k!}-\sum_{k=0}^{n}\frac{a^k}{k!}|<\frac{\epsilon}{3}$$

$$|\sum_{k=0}^{m}\frac{b^k}{k!}-\sum_{k=0}^{n}\frac{b^k}{k!}|<\frac{\epsilon}{3}$$

$$|\sum_{k=0}^{n}\frac{b^k}{k!}-\sum_{k=0}^{n}\frac{a^k}{k!}|<\frac{\epsilon}{3}.$$

From the first two inequalities we get

$$|\exp(a)-\sum_{k=0}^{n}\frac{a^k}{k!}|\leq\frac{\epsilon}{3}$$

$$|\exp(b)-\sum_{k=0}^{n}\frac{b^k}{k!}|\leq\frac{\epsilon}{3}.$$

Then

$$|\exp(b)-\exp(a)|<\frac{\epsilon}{3}+\frac{\epsilon}{3}+\frac{\epsilon}{3}=\epsilon.$$

EXERCISE 38.10. Let $S \subset \{0,1\}^{\mathbb{N}}$ be the set of all sequences (a_n) such that there exist N with $a_n = 1$ for all $n \geq N$. Prove that the map

$$\{0,1\}^{\mathbb{N}} \backslash S \to \mathbb{R}, \quad (a_n) \mapsto \sum_{n=1}^{\infty} \frac{a_n}{2^n}$$

is (well defined and) injective. Conclude that $\mathbb{R}$ is not countable.

EXERCISE 38.11. Prove that there exist transcendental numbers in $\mathbb{R}$. Hint: $\mathbb{R}$ is uncountable whereas the set of algebraic numbers is countable; cf. Exercise 32.20. This is Cantor's proof of existence of transcendental numbers.

Real analysis (analysis of sequences, continuity, and other concepts of calculus like differentiation and integration of functions on $\mathbb{R}$) can be extended to complex analysis. Indeed we have:

DEFINITION 38.12. A sequence (z_n) in $\mathbb{C}$ is convergent to $z_0 \in \mathbb{C}$ if for any real number $\epsilon > 0$ there exists an integer N such that for all $n \geq N$ we have $|z_n - z_0| < \epsilon$. We write $z_n \to z_0$ and we say z_0 is the limit of (z_n). A sequence is called convergent if there exists $z \in \mathbb{C}$ such that the sequence converges to z. A sequence is called divergent if it is not convergent.

EXERCISE 38.13. Let (z_n) be a sequence in $\mathbb{C}$ and let

$$z_n = a_n + b_n i,$$

$a_n, b_n \in \mathbb{R}$. Let $z_0 = a_0 + b_0 i$. Prove that $z_n \to z_0$ if and only if $a_n \to a_0$ and $b_n \to b_0$.

DEFINITION 38.14. A sequence (z_n) in $\mathbb{C}$ is Cauchy if for any real $\epsilon > 0$ there exists an integer N such that for all integers $m, n \geq N$ we have $|z_n - z_m| < \epsilon$.

EXERCISE 38.15. Prove that a sequence in $\mathbb{C}$ is convergent if and only if it is Cauchy.

EXERCISE 38.16. Prove that:
1) The series

$$\sum_{n=0}^{\infty} \frac{z^n}{n!}$$

is convergent for all $z \in \mathbb{C}$; its limit is denoted by $e^z = \exp(z)$.
2) $\exp(z + w) = \exp(z)\exp(w)$ for all $z, w \in \mathbb{C}$.
3) $\overline{\exp(z)} = \exp(\bar{z})$ for all $z \in \mathbb{C}$.
4) $|\exp(it)| = 1$ for all $t \in \mathbb{R}$.
5) The map

$$\mathbb{C} \to \mathbb{C}, \quad z \mapsto \exp(z),$$

is continuous. This map is called the (complex) exponential map.

There is a version of the above theory in p-adic analysis (which is crucial to number theory). Recall the ring of p-adic numbers $\mathbb{Z}_p$ whose elements are denoted by $[a_n]$.

DEFINITION 38.17. Say that p^e divides $\alpha = [a_n]$ if there exists $\beta = [b_n]$ such that $[a_n] = [p^e b_n]$; write $p^e | \alpha$. For any $0 \neq \alpha = [a_n] \in \mathbb{Z}_p$ let $v = v(\alpha)$ be the unique integer such that $p^n | a_n$ for $n \leq v$ and $p^{v+1} \nmid a_{v+1}$. Then define the norm of α by the formula $|\alpha| = p^{-v(\alpha)}$. We also set $|0| = 0$.

EXERCISE 38.18. Prove that if $\alpha = [a_n]$ and $\beta = [b_n]$ then
$$|\alpha + \beta| \leq \max\{|\alpha|, |\beta|\}.$$

DEFINITION 38.19. Consider a sequence $[a_{n1}], [a_{n2}], [a_{n3}], \ldots$ of elements in $\mathbb{Z}_p$ which for simplicity we denote by $\alpha_1, \alpha_2, \alpha_3, \ldots$.
1) $\alpha_1, \alpha_2, \alpha_3, \ldots$ is called a Cauchy sequence if for any real (or, equivalently, rational) $\epsilon > 0$ there exists an integer N such that for all $m, m' \geq N$ we have $|\alpha_m - \alpha_{m'}| \leq \epsilon$.
2) We say that $\alpha_1, \alpha_2, \alpha_3, \ldots$ converges to some $\alpha_0 \in \mathbb{Z}_p$ if for any real (or, equivalently, rational) $\epsilon > 0$ there exists an integer N such that for all $m \geq N$ we have $|\alpha_m - \alpha_0| \leq \epsilon$. We say α_0 is the limit of (α_n) and we write $\alpha_n \to \alpha_0$.

EXERCISE 38.20. Prove that a sequence in $\mathbb{Z}_p$ is convergent if and only if it is Cauchy.

The following is in deep contrast with the case of $\mathbb{R}$:

EXERCISE 38.21. Prove that if (α_n) is a sequence in $\mathbb{Z}_p$ with $\alpha_n \to 0$ then the sequence $s_n = \sum_{k=1}^{n} \alpha_k$ is convergent in $\mathbb{Z}_p$; the limit of the latter is denoted by $\sum_{n=1}^{\infty} \alpha_n$.

EXERCISE 38.22.
1) Prove that $\sum_{n=1}^{\infty} p^{n-1}$ is the inverse of $1 - p$ in $\mathbb{Z}_p$.
2) Prove that if $\alpha \in \mathbb{Z}_p$ has $|\alpha| = 1$ then α is invertible in $\mathbb{Z}_p$. Hint: Use the fact that if p does not divide an integer $a \in \mathbb{Z}$ then there exist integers $m, n \in \mathbb{A}$ such that $ma + np = 1$; then use 1) above.
3) Prove that for all $n \geq 1$ and all $a \in \mathbb{Z}_p$ with $|a| < 1$ there exists an element of $\mathbb{Z}_p$ denoted by $\frac{a^n}{n!}$ such that $(n!) \cdot \frac{a^n}{n!} = a^n$. Hint: Use 2) above.
4) Prove that $\sum_{n=1}^{\infty} \frac{a^n}{n!}$ is convergent in $\mathbb{Z}_p$ for all $a \in \mathbb{Z}_p$ with $|a| < 1$. One denotes the limit by $\exp_p(a)$.

CHAPTER 39

Trigonometry

Trigonometry arose long before calculus mainly motivated by geometry and astronomy. A rigorous approach to trigonometry requires some elements of analysis that we already covered and hence can be used in what follows. We will define the functions sin and cos and also the number π.

DEFINITION 39.1. For all $t \in \mathbb{R}$ define $\cos t, \sin t \in \mathbb{R}$ as being the unique real numbers such that

$$\exp(it) = \cos t + i \sin t.$$

(This is called Euler's formula but here this is a definition and not a theorem.)

EXERCISE 39.2. Prove the following equalities:
1) $\cos(t_1 + t_2) = \cos t_1 \cos t_2 - \sin t_1 \sin t_2$;
2) $\sin(t_1 + t_2) = \sin t_1 \cos t_2 + \cos t_1 \sin t_2$.

EXERCISE 39.3. Prove that the map $f : \mathbb{R} \to SO_2(\mathbb{R})$ defined by

$$f(t) = \begin{pmatrix} \cot t & \sin t \\ -\sin t & \cos t \end{pmatrix}$$

is a group homomorphism.

EXERCISE 39.4. Prove that if H is a closed subgroup of $\mathbb{R}$ and $H \neq \mathbb{R}$, $H \neq \{0\}$ then there exists a unique $T \in \mathbb{R}$, $T > 0$, such that

$$H = \{nT; n \in \mathbb{Z}\}.$$

Hint: One first shows that if T is the infimum of the set

$$\{a \in H; a > 0\}$$

then $T \neq 0$. In order to check this assume $T = 0$ and seek a contradiction. Indeed from $T = 0$ we get that there exists a sequence (a_n) with $a_n \in H$, and $a_n \to 0$. Deduce from this and the fact that H is closed that $G = \mathbb{R}$, a contradiction. Finally one shows that $H = \{nT; n \in \mathbb{Z}\}$ using an argument similar to the one used to prove Proposition 25.6.

EXERCISE 39.5. Prove that the map

$$F : \mathbb{R} \to \mathbb{C}^\times, F(t) = \exp(it)$$

is non-constant and non-injective. Conclude that there exists a unique real number $\pi \in \mathbb{R}$, $\pi > 0$, such that

$$Ker \; F = \{2n\pi; n \in \mathbb{Z}\}.$$

(This is our definition of the number π. In particular, by this very definition one gets $\exp(\pi i) + 1 = 0$ which is a celebrated formula of Euler; for us this is a trivial consequence of our definition of π.)

CHAPTER 40

Integrality

A key concept in number theory is that of an algebraic integer. It emerged through the work of Gauss, Dirichlet, Eisenstein, Kummer, Kronecker, Hilbert, and others. It is a generalization of the notion of (usual) integer; and many theorems about algebraic integers have, as consequences, theorems about the usual integers. A typical example of this will be the proof of the Quadratic Reciprocity Law of Gauss.

DEFINITION 40.1. A complex number $u \in \mathbb{C}$ is called an algebraic integer if there exists a monic polynomial $F \in \mathbb{Z}[x]$ such that $F(u) = 0$.

EXAMPLE 40.2. $\sqrt{-7} := \sqrt{7}i \in \mathbb{C}$ is an algebraic integer because it is a root of $F(x) = x^2 + 7$. N.B. Not all algebraic integers can be obtained from rational numbers by iterating the operations of addition, multiplication, and taking radicals of various orders; in order to prove the existence of algebriac integers that cannot be obtained in this way one needs "Galois theory."

DEFINITION 40.3. A subset $\mathcal{O} \subset \mathbb{C}$ is called an order if:
1) $1 \in \mathcal{O}$;
2) $u, v \in \mathcal{O}$ implies $u + v, uv, -u \in \mathcal{O}$;
3) There exist $u_1, ..., u_n \in \mathcal{O}$ such that
$$\mathcal{O} = \{m_1 u_1 + ... + m_n u_n; m_1, ..., m_n \in \mathbb{Z}\}.$$

The word *order* above has nothing to do with the word *order* which was earlier used in the phrases *order relation*, or *order of an element in a group*, or *order of a group*.

REMARK 40.4. Conditions 1 and 2 imply that $\mathcal{O}$ is a ring with respect to $+$ and $\times$.

EXERCISE 40.5. Prove that the sets
$$\mathbb{Z}[i], \quad \{a + 2b\sqrt{-7}; a, b, \in \mathbb{Z}\}, \quad \{a + 2b\sqrt{7}; a, b, \in \mathbb{Z}\}$$
are orders. Draw pictures of these sets.

PROPOSITION 40.6. *A complex number is an algebraic integer if and only if it is contained in an order.*

Proof. If u is an algebraic integer, root of a monic polynomial in $\mathbb{Z}[x]$ of degree n then u is contained in the order
$$\mathcal{O} := \{c_0 + c_1 u + ... + c_{n-1} u^{n-1}; c_0, ..., c_{n-1} \in \mathbb{Z}\}.$$
Conversely assume u is contained in the order
$$\mathcal{O} = \{m_1 u_1 + ... + m_n u_n; m_1, ..., m_n \in \mathbb{Z}\}.$$

Then for all $i = 1, ..., n$ we can write

$$uu_i = \sum_{j=1}^{n} m_{ij} u_j$$

with $m_{ij} \in \mathbb{Z}$. Set $a_{ij} = \delta_{ij} u - m_{ij}$ where δ_{ij} is the Kronecker symbol, i.e., 1 or 0 according as $i = j$ or $i \neq j$. Let $A = (a_{ij})$ be the matrix with entries a_{ij} and let b be the column vector with entries u_i. Since $Ab = 0$ and $b \neq 0$ it follows that A is not invertible hence $\det(A) = 0$. But $\det(A)$ is easily seen to have the form

$$\det(A) = u^n + a_1 u^{n-1} + ... + a_{n-1} u + a_n$$

with $a_k \in \mathbb{Z}$ so u is an algebraic integer and we are done. $\square$

PROPOSITION 40.7. *If u and v are algebraic integers then $u + v, uv, -u$ are also algebraic integers.*

Proof. Assume u belongs to the order

$$\{a_1 u_1 + ... + a_n u_n; a_1, ..., a_n \in \mathbb{Z}\}$$

and v belongs to the order

$$\{b_1 v_1 + ... + b_m v_m; b_1, ..., b_m \in \mathbb{Z}\}.$$

Then $u + v, uv, -u$ belong to the set

$$\mathcal{O} = \left\{ \sum_{i=1}^{n} \sum_{j=1}^{m} c_{ij} u_i v_j; c_{ij} \in \mathbb{Z} \right\};$$

but $\mathcal{O}$ is clearly an order. $\square$

DEFINITION 40.8. Denote by $\overline{\mathbb{Z}} \subset \mathbb{C}$ the set of all algebraic integers.

REMARK 40.9. By Proposition 40.7 $\overline{\mathbb{Z}}$ is a ring with respect to $+$ and $\times$.

PROPOSITION 40.10. *A rational number which is also an algebraic integer must be an integer. In other words $\overline{\mathbb{Z}} \cap \mathbb{Q} = \mathbb{Z}$.*

Proof. Assume $\frac{a}{b} \in \mathbb{Q}$ is an algebraic integer,

$$\left(\frac{a}{b}\right)^n + a_1 \left(\frac{a}{b}\right)^{n-1} + ... + a_n = 0$$

with $a_1, ..., a_n \in \mathbb{Z}$. Hence

$$a^n + a_1 a^{n-1} b + ... + a_n b^n = 0.$$

Assume $\frac{a}{b} \notin \mathbb{Z}$. Then there exists a prime $p \in \mathbb{Z}$ with $p|b$ and $p \nmid a$. But by the last equation if $p|b$ then $p|a^n$ hence $p|a$, a contradiction. $\square$

EXERCISE 40.11. Find an order containing $\sqrt{3} + \sqrt{7}$. Find a similar example involving cubic roots.

EXERCISE 40.12. Find a monic polynomial $f(x)$ in $\mathbb{Z}[x]$ such that $f(\sqrt{3}+\sqrt{7}) = 0$. Find a similar example involving cubic roots.

EXERCISE 40.13. Let $n \geq 2$ be an integer. Prove that there exists a complex number $\zeta_n \in \mathbb{C}$ such that $\zeta_n^n = 1$ and $\zeta_n^k \neq 1$ for $1 \leq k \leq n - 1$. (Note that ζ_n is then an algebraic integer, $\zeta_n \in \overline{\mathbb{Z}}$.)

Hint: One way to find ζ_n is to take

$$\zeta_n = \exp(\frac{2\pi i}{n}).$$

Alternatively, if we assume the Fundamental Theorem of Algebra (Theorem 32.13) then one can take ζ_n to be an appropriate root of the polynomial $x^n - 1$.

CHAPTER 41

Reciprocity

We prove here the Quadratic Reciprocity Law of Gauss. This is one of the most powerful (and mysterious) theorems in elementary number theory; much of the number theory after Gauss was motivated by the attempt to "understand" and generalize this theorem. Cf. work of Eisenstein, Kummer, Hilbert, Artin, Tate, and many others.

Let p be a prime $\neq 2$.

DEFINITION 41.1. If a is any integer define the Legendre symbol

$$\left(\frac{a}{p}\right) = N_p(x^2 = a) - 1,$$

i.e., the Legendre symbol is $-1, 0, 1$ according as $x^2 \equiv a \ (mod \ p)$ has two solutions, one solution (this is the case if and only if $p|a$), or no solution mod p, respectively.

EXERCISE 41.2. Prove that

$$\left(\frac{ab}{p}\right) = \left(\frac{a}{p}\right)\left(\frac{b}{p}\right).$$

LEMMA 41.3. *(Euler)*

$$\left(\frac{a}{p}\right) \equiv a^{\frac{p-1}{2}} \ (mod \ p).$$

Proof. Let $\bar{a} \in \mathbb{F}_p^\times$ denote the image of any $a \in \mathbb{Z}$ which is not divisible by p. Then the following maps

$$\bar{a} \mapsto \bar{a}^{\frac{p-1}{2}}, \quad \bar{a} \mapsto \overline{\left(\frac{a}{p}\right)}$$

are group homomorphisms $\mathbb{F}_p^\times \to \{\bar{1}, -\bar{1}\}$. Since $\mathbb{F}_p^\times$ is cyclic generated by $\bar{g}$ with g a primitive root of unity it is enough to prove that

$$\left(\frac{g}{p}\right) = -1 \quad \text{and} \quad \bar{g}^{\frac{p-1}{2}} = -1.$$

The second follows from the fact that g is a primitive root. Assume now $\left(\frac{g}{p}\right) = 1$ and seek a contradiction. Indeed from our assumption it follows that any element in $\mathbb{F}_p^\times$ is a square. This implies that the homomorphism $\mathbb{F}_p^\times \to \mathbb{F}_p^\times$, $\bar{a} \mapsto \bar{a}^2$ is surjective hence injective. But this is a contradiction because the kernel of the latter homomorphism consists of $\bar{1}$ and $-\bar{1}$. This ends the proof. $\square$

COROLLARY 41.4. *(Euler)* $\left(\frac{-1}{p}\right) = 1$ *if and only if* $p \equiv 1 \ (mod \ 4)$.

LEMMA 41.5. *(Euler)* $\left(\frac{2}{p}\right) = 1$ *if and only if* $p \equiv 1, 7 \ (mod \ 8)$.

Proof. First we claim that $\left(\frac{2}{p}\right) = (-1)^\mu$ where μ is the number of integers in the set $\{2, 4, 6, ..., p-1\}$ congruent mod p to a negative integer between $-\frac{p}{2}$ and $\frac{p}{2}$. Indeed let $r_1, ..., r_{\frac{p-1}{2}}$ be the integers between $-\frac{p}{2}$ and $\frac{p}{2}$ that are congruent mod p to $2, 4, 6, ..., p-1$. Then it is easy to check that

$$\{|r_1|, ..., |r_{\frac{p-1}{2}}|\} = \{1, 2, 3, ..., \frac{p-1}{2}\},$$

where $|r|$ is the absolute value of r. Taking products we get

$$1 \times 2 \times 3 \times ... \times \frac{p-1}{2} \equiv (-1)^\mu \times 2^{\frac{p-1}{2}} \times 1 \times 2 \times 3 \times ... \times \frac{p-1}{2} \quad (mod\ p)$$

which proves our claim.

Now note that if an integer a between $-\frac{p}{2}$ and 0 is congruent mod p to one of the numbers $2, 4, 6, ..., p-1$ then $2x \equiv a\ (mod\ p)$ for some $x \in \{1, 2, 3, ..., \frac{p-1}{2}\}$. Writing $a = 2x + mp$ we get $-\frac{p}{2} < 2x + mp < 0$ hence $\frac{p}{2} < 2x + (m+1)p < p$ which forces $m = -1$ hence $\frac{p}{2} < 2x < p$. Conversely if the latter holds then $a = 2x - p$ is between $-\frac{p}{2}$ and 0. So if $p = 8k + r$, $0 \le r < 7$, we have

$$\mu = |\{x \in \mathbb{Z}; \frac{p}{2} < 2x < p\}|$$

$$= |\{x \in \mathbb{Z}; \frac{p}{4} < x < \frac{p}{2}\}|$$

$$= |\{x \in \mathbb{Z}; 2k + \frac{r}{4} < x < 4k + \frac{r}{2}\}|$$

$$= |\{x \in \mathbb{Z}; \frac{r}{4} < x < 2k + \frac{r}{2}\}|$$

and we conclude by inspecting the values $r = 1, 3, 5, 7$.   □

EXERCISE 41.6.

$$\sum_{a=1}^{p-1} \left(\frac{a}{p}\right) = 0.$$

Hint: Use the fact that the image of the homomorphism $\mathbb{F}_p^\times \to \mathbb{F}_p^\times$, $\bar{a} \mapsto \bar{a}^2$ has $\frac{p-1}{2}$ elements.

The main result pertaining to the Legendre symbol is the following theorem of Gauss:

THEOREM 41.7. *(Quadratic Reciprocity Law) For any two distinct primes p and q different from 2 we have*

$$\left(\frac{p}{q}\right)\left(\frac{q}{p}\right) = (-1)^{\frac{p-1}{2}\frac{q-1}{2}}.$$

We will prove this presently.

Theorem 41.7 plus Lemma 41.5 imply:

COROLLARY 41.8. *If $a \in \mathbb{N}$ and p_1, p_2 are primes such that*

$$p_1 \equiv p_2\ (mod\ 4a)$$

then

$$N_{p_1}(x^2 = a) = N_{p_2}(x^2 = a).$$

In other words if we fix the polynomial $f(x) = x^2 - a$ then there exists an integer N (in our case $N = 4a$) such that for any p the value of $N_p(f)$ only depends on the image of p in $\mathbb{Z}/N\mathbb{Z}$. Such a statement fails, in general, for polynomials f of arbitrary degree (although there are examples of higher degree for which such a statement holds).

Let $\zeta_p \in \mathbb{C}$, $\zeta_p^p = 1$, $\zeta_p \neq 1$. We will need the following:

EXERCISE 41.9. Prove that if c is an integer then

$$\sum_{b=1}^{p-1} (\zeta_p^c)^b$$

equals $p - 1$ or -1 according as $p | c$ or $p \nmid c$.

DEFINITION 41.10. Define the Gauss sum

$$G = \sum_{a=1}^{p-1} \left(\frac{a}{p} \right) \zeta_p^a \in \overline{\mathbb{Z}}.$$

LEMMA 41.11. *(Gauss)*

$$G^2 = (-1)^{\frac{p-1}{2}} p.$$

Proof. We have

$$G^2 = \sum_{a=1}^{p-1} \sum_{b=1}^{p-1} \left(\frac{ab}{p} \right) \zeta_p^{a+b}.$$

If (a, b) runs through the set of indices of the above sum then clearly $(\overline{a}, \overline{ab})$ runs through $\mathbb{F}_p^{\times} \times \mathbb{F}_p^{\times}$ so substituting a by ab the above sum equals

$$\sum_{a=1}^{p-1} \sum_{b=1}^{p-1} \left(\frac{ab^2}{p} \right) \zeta_p^{ab+b} = \sum_{a=1}^{p-1} \left(\frac{a}{p} \right) \sum_{b=1}^{p-1} (\zeta_p^{a+1})^b.$$

In view of Exercises 41.9 and 41.6 the above sum equals

$$\left(\frac{-1}{p} \right) (p - 1) - \sum_{a=1}^{p-2} \left(\frac{a}{p} \right) = \left(\frac{-1}{p} \right) p$$

and we are done by Lemma 41.3.   □

DEFINITION 41.12. For $u, v \in \overline{\mathbb{Z}}$ and q a prime in $\mathbb{Z}$ let us write $u \equiv v \pmod{q}$ in $\overline{\mathbb{Z}}$ if there exists $w \in \overline{\mathbb{Z}}$ such that $qw = v - u$.

EXERCISE 41.13. Prove that if $u \equiv v \pmod{q}$ in $\overline{\mathbb{Z}}$ and $u, v \in \mathbb{Z}$ then $u \equiv v \pmod{q}$ in $\mathbb{Z}$. Hint: This follows directly from Proposition 40.10.

EXERCISE 41.14. (Freshman's Dream) Prove that

$$(u_1 + ... + u_n)^p \equiv u_1^p + ... + u_n^p \pmod{p} \text{ in } \overline{\mathbb{Z}}$$

for $u_1, ..., u_n \in \overline{\mathbb{Z}}$ and p a prime in $\mathbb{Z}$.

Hint: Use induction on n plus the fact that p divides all the binomial coefficients $\binom{p}{k}$ for $1 \leq k \leq p - 1$.

Proof of Theorem 41.7. By Lemma 41.11 and then Lemma 41.3

$$G^q = G(G^2)^{\frac{q-1}{2}} = G(-1)^{\frac{p-1}{2}\frac{q-1}{2}} p^{\frac{q-1}{2}} \equiv G(-1)^{\frac{p-1}{2}\frac{q-1}{2}} \left(\frac{p}{q}\right) \ (mod\ q) \text{ in } \overline{\mathbb{Z}}.$$

On the other hand by "Freshman's Dream" we get

$$G^q = \left(\sum_{a=1}^{p-1} \left(\frac{a}{p}\right) \zeta_p^a\right)^q \equiv \sum_{a=1}^{p-1} \left(\frac{a}{p}\right)^q \zeta_p^{aq} = \sum_{a=1}^{p-1} \left(\frac{a}{p}\right) \zeta_p^{aq}$$

$$= \left(\frac{q}{p}\right) \sum_{a=1}^{p-1} \left(\frac{aq}{p}\right) \zeta_p^{aq} = \left(\frac{q}{p}\right) G \ (mod\ q) \text{ in } \overline{\mathbb{Z}}.$$

The two expressions of G^q above give

$$G(-1)^{\frac{p-1}{2}\frac{q-1}{2}} \left(\frac{p}{q}\right) \equiv \left(\frac{q}{p}\right) G \ (mod\ q) \text{ in } \overline{\mathbb{Z}}.$$

Assume

$$(-1)^{\frac{p-1}{2}\frac{q-1}{2}} \left(\frac{p}{q}\right) \neq \left(\frac{q}{p}\right),$$

and let us derive a contradiction. Since the two numbers above are ± 1 we get that one is 1 and the other is -1 so we get

$$G \equiv -G \quad (mod\ q) \text{ in } \overline{\mathbb{Z}}$$

hence

$$2G \equiv 0 \quad (mod\ q) \text{ in } \overline{\mathbb{Z}}.$$

Squaring and using Lemma 41.11 we get

$$4p \equiv 0 \quad (mod\ q) \text{ in } \overline{\mathbb{Z}}$$

and hence, by Exercise 41.13,

$$4p \equiv 0 \quad (mod\ q) \text{ in } \mathbb{Z}$$

which is a contradiction. □

CHAPTER 42

Calculus

Calculus was invented by Newton and Leibniz, motivated by problems in mechanics and analytic geometry. The main concept of calculus is that of derivative of a function which we briefly review here.

DEFINITION 42.1. Let $F : \mathbb{R} \to \mathbb{R}$ be a map and $a_0 \in \mathbb{R}$. We say F is differentiable at a_0 if there exists a real number (denoted by) $F'(a_0) \in \mathbb{R}$ such that for any sequence $a_n \to a_0$ with $a_n \neq a_0$ we have

$$\frac{F(a_n) - F(a_0)}{a_n - a_0} \to F'(a_0).$$

EXERCISE 42.2. Prove that if F is differentiable at $a_0 \in \mathbb{R}$ then it is continuous at a_0.

EXERCISE 42.3. Prove that if F is a constant function (i.e., $F(x) = F(y)$ for all $x, y \in \mathbb{R}$) then F is differentiable at any a and $F'(a) = 0$.

DEFINITION 42.4. We say $F : \mathbb{R} \to \mathbb{R}$ is differentiable if F is differentiable at any $a \in \mathbb{R}$. If this is the case the map $a \mapsto F'(a)$ is called the derivative of F and is denoted by $F' = \frac{dF}{dx} : \mathbb{R} \to \mathbb{R}$. If F' is differentiable we say F is twice differentiable and F'' is called the second derivative of F. One similarly defines what it means for F to be n times differentiable ($n \in \mathbb{N}$). We say F is infinitely differentiable (or smooth) if it is n times differentiable for any $n \in \mathbb{N}$. One denotes by $C^\infty(\mathbb{R})$ the set of smooth functions. We denote by $D : C^\infty(\mathbb{R}) \to C^\infty(\mathbb{R})$ the map $D(F) = F'$.

EXERCISE 42.5. Prove that for any $F, G \in C^\infty(\mathbb{R})$ we have $F+G, F \cdot G \in C^\infty(\mathbb{R})$ and
  1) $D(F + G) = D(F) + D(G)$ (additivity);
  2) $D(F \cdot G) = F \cdot D(G) + G \cdot D(F)$ (Leibniz rule);
here $F + G, F \cdot G$ are the pointwise addition and multiplication of F and G. In particular $C^\infty(\mathbb{R})$ is a ring with respect to $+$ and $\cdot$; 0 and 1 are the functions $0(x) = 0$ and $1(x) = 1$.

EXERCISE 42.6. Prove that any polynomial function $F : \mathbb{R} \to \mathbb{R}$ is smooth and

$$F(x) = \sum_{k=0}^{n} a_n x^n \Rightarrow F'(x) = \sum_{k=0}^{n} n a_n x^{n-1}.$$

Hint: It is enough to look at $F(x) = x^k$. In this case

$$\frac{a_n^k - a_0^k}{a_n - a_0} = a_n^{k-1} + a_n^{k-2} a_0 + \ldots + a_0^{k-1} \to k a_0^{k-1}.$$

EXERCISE 42.7. Prove that $F(x) = \exp(x)$ is differentiable and $F'(x) = \exp(x)$. Hence F is smooth.

EXERCISE 42.8. Prove that $F(x) = \sin(x)$ is differentiable and $F'(x) = \cos(x)$. Prove that $G(x) = \cos(x)$ is differentiable and $G'(x) = -\sin(x)$. Hence F and G are smooth.

EXERCISE 42.9. (Chain rule) Prove that if $F, G \in C^\infty(\mathbb{R})$ then $F \circ G \in C^\infty(\mathbb{R})$ and
$$D(F \circ G) = (D(F) \circ G) \cdot D(G).$$
(Here, as usual, $\circ$ denotes composition.)

More generally one can define derivatives of functions of several variables as follows:

DEFINITION 42.10. Let $F : \mathbb{R}^n \to \mathbb{R}$ be a function. Let $a = (a_1, ..., a_n) \in \mathbb{R}^n$ and define $F_i : \mathbb{R} \to \mathbb{R}$ by
$$F_i(x) = F(a_1, ..., a_{i-1}, x, a_{i+1}, ..., a_n)$$
(with the obvious adjustment if $i = 1$ or $i = n$). We say that F is differentiable with respect to x_i at a if F_i is differentiable at a_i; in this case we define
$$\frac{\partial F}{\partial x_i}(a) = F_i'(a_i).$$
We say that F is differentiable with respect to x_i if it is differentiable with respect to x_i at any $a \in \mathbb{R}^n$. For such a function we have a well defined function $\frac{\partial F}{\partial x_i} : \mathbb{R}^n \to \mathbb{R}$ which is also denoted by $D_i F$. We say that F is infinitely differentiable (or smooth) if F is differentiable, each $D_i F$ is differentiable, each $D_i D_j F$ is differentiable, each $D_i D_j D_k F$ is differentiable, etc. We denote by $C^\infty(\mathbb{R}^n)$ the set of smooth functions; it is a ring with respect to pointwise addition and multiplication.

DEFINITION 42.11. Let $P \in C^\infty(\mathbb{R}^{r+2})$. An equation of the form
$$P\left(x, F(x), \frac{dF}{dx}(x), \frac{d^2F}{dx^2}(x), ..., \frac{d^rF}{dx^r}(x)\right) = 0$$
is called a differential equation. Here $F \in C^\infty(\mathbb{R})$ is an unknown function and one defines $\frac{d^iF}{dx^i} = D^{i+1}(F) = D(D^i(F))$.

The study of differential equations has numerous applications within mathematics (e.g., geometry) as well as natural sciences (e.g., physics).

EXAMPLE 42.12. Here is a random example of a differential equation:
$$\exp\left(\left(\frac{d^2F}{dx^2}\right)^5\right) - x^3\left(\frac{dF}{dx}\right)\left(\frac{d^3F}{dx^3}\right) - x^5F^6 = 0.$$

The additivity and the Leibniz rule have an algebraic flavor. This suggests the following:

DEFINITION 42.13. Let R be a commutative unital ring. A map $D : R \to R$ is called a derivation if
1) $D(a + b) = D(a) + D(b)$ (additivity);
2) $D(a \cdot b) = a \cdot D(b) + b \cdot D(a)$ (Leibniz rule).

EXAMPLE 42.14. $D : C^\infty(\mathbb{R}) \to C^\infty(\mathbb{R})$ is a derivation.

EXERCISE 42.15. Prove that any derivation $D : \mathbb{Z} \to \mathbb{Z}$ is identically 0 i.e., $D(x) = 0$ for all $x \in \mathbb{Z}$.

Hint: By additivity $D(0) = 0$ and $D(-n) = -D(n)$. So it is enough to show $D(n) = 0$ for $n \in \mathbb{N}$. Proceed by induction on n. For the case $n = 1$, by the Leibniz rule,
$$D(1) = D(1 \cdot 1) = 1 \cdot D(1) + 1 \cdot D(1) = 2 \cdot D(1)$$
hence $D(1) = 0$. The induction step follows by additivity.

REMARK 42.16. Exercise 42.15 shows that there is no naive analogue of calculus in which rings of functions such as $C^\infty(\mathbb{R})$ are replaced by rings of numbers such as $\mathbb{Z}$. An analogue of calculus for $\mathbb{Z}$ is, however, considered desirable for the purposes of number theory. Such a theory has been developed. (Cf. A. Buium, *Arithmetic Differential Equations*, Mathematical Surveys and Monographs 118, American Mathematical Society, 2005.) In that theory the analogue of x is a fixed prime p and the analogue of the derivation $D = \frac{d}{dx} : C^\infty(\mathbb{R}) \to C^\infty(\mathbb{R})$ is the operator
$$\frac{d}{dp} : \mathbb{Z} \to \mathbb{Z}, \quad \frac{dx}{dp} = \frac{x - x^p}{p}$$
which is well defined by Fermat's Little Theorem. For example,
$$\frac{d4}{d5} = \frac{4 - 4^5}{5} = -204.$$
In order for the theory to achieve its full strength one needs to replace $\mathbb{Z}$ by orders of the form
$$\mathcal{O} = \mathbb{Z}[\zeta_n] = \{\sum n_i \zeta_n^i ; n_i \in \mathbb{Z}\}$$
where $p \nmid n$; also one needs to consider a generalization of the operator $\frac{d}{dp}$ above given by
$$\frac{d}{dp} : \mathcal{O} \to \mathcal{O}, \quad \frac{dx}{dp} = \frac{\phi(x) - x^p}{p}$$
where $\phi : \mathcal{O} \to \mathcal{O}$ is the unique ring isomorphism such that $\phi(\zeta_n) = \zeta_n^p$ (this ϕ exists by elementary Galois theory); one also needs to consider analogues of algebraic differential equations of the form
$$P\left(x, \frac{dx}{dp}, \frac{d^2x}{dp^2}, ..., \frac{d^r x}{dp^r}\right) = 0$$
in which x and P are elements in the p-adic completions
$$\widehat{\mathcal{O}} \quad \text{and} \quad \widehat{\mathcal{O}[y_0, ..., y_r]}$$
of $\mathcal{O}$ and $\mathcal{O}[y_0, ..., y_r]$, respectively; cf. 26.8. We will not pursue this here. Let us just note that the operators $\frac{d}{dp}$ are not additive and do not satisfy the Leibniz rule; it is therefore surprising that the theory of arithmetic differential equations alluded to above closely parallels the classical theory of differential equations; it also has purely number theoretic applications.

CHAPTER 43

Metamodels

In this chapter we briefly discuss the problem of translating various theories into set theory. One can give the following metadefinition which is *not standard* but expresses a standard idea about the interaction between mathematics and various fields of knowledge (including mathematics itself). Later, when we discuss mathematical logic, we will introduce a rather different (but related) concept, namely that of a *model*.

METADEFINITION 43.1. Let T be a theory in a language L with specific axioms $A, B, \dots$. Assume in addition that one may be given a "domain" formula $D(x)$ in L_{set}^f with one free variable x. (We allow that D be not present.) By a mathematical model with domain D (or simply a metamodel) for T we mean data $\tilde{L}, \tilde{T}, \tilde{\ }$ consisting of

i) a language $\tilde{L}$ obtained from the language of set theory L_{set} by adding various constants and predicates plus definitions for each of the predicates;

ii) a D-translation $\tilde{\ }$ of L into the language $\tilde{L}$; and

iii) a theory $\tilde{T}$ in $\tilde{L}$ such that:

1) $\tilde{T}$ contains T_{set};
2) $\tilde{T}$ contains the translations $\tilde{A}, \tilde{B}, \dots$ of the specific axioms $A, B, \dots$ of T;
3) $\tilde{T}$ contains $D(\tilde{t})$ for all terms t without free variables in L.

REMARK 43.2. It is easy to check that the D-translations of all background axioms of T are contained in $\tilde{T}$; therefore the D-translation of any theorem in T will be a theorem in $\tilde{T}$.

REMARK 43.3. If D in the above metadefinition is not present then D-translations are simply translations and condition 3) above can be dropped.

REMARK 43.4. The concept of metamodel introduced above is completely different from (but also an inspiration for) the concept of model to be introduced later in the chapter on models. Metamodels consist of texts; models in the chapter on models are sets.

EXAMPLE 43.5. We illustrate metamodels by considering the "toy" theory T of gravitation in Example 11.19. This example will not involve any "domain formula D." Let L be the language in Example 11.19; recall that it contains constants $S, E, M, R, 1, \pi, g$, relational predicates $p, c, n, \circ, f$, and functional symbols $d, a, T, :$, $\times$. We need to first specify a language $\tilde{L}$. We let $\tilde{L}$ be the language obtained from L_{set} by adding symbols $\tilde{S}, \dots, \tilde{T}$ (all the symbols above with a tilde on top) The translation of L into $\tilde{L}$ is then just "putting $\tilde{\ }$ on top of the symbols." We let $\mathcal{F}$ be a fixed subset of the set of smooth maps $\mathbb{R} \to \mathbb{R}^3$ (i.e., maps whose components $x_1, x_2, x_3 : \mathbb{R} \to \mathbb{R}$ are smooth). Then we consider definitions such as:

1) $\tilde{\pi} = 314 : 100$, $\tilde{g} = 981,...$;

2) $\forall x(\tilde{n}(x) \leftrightarrow ((x \in \mathbb{R})) \wedge (x > 0))$;

3) $\forall x \forall y(\tilde{f}(x, y) \leftrightarrow ((x \in \mathcal{F}) \wedge (y \in \mathbb{R})))$;

4) $\tilde{o}$ expresses movement in a circle (using the sin and cos functions);

5) $\tilde{a}(x) = \sqrt{(x_1'')^2 + (x_2'')^2 + (x_3'')^2}$, where x_i'' is the second derivative of x_i, etc.

Finally, $\tilde{T}$ is the theory with specific axioms, those of T_{set} to which one adds the translations in $\tilde{L}$ of the axioms in T in Example 11.19. The above yields a metamodel of T.

EXERCISE 43.6. Attempt to make the above more explicit and to complete the definitions above. Analyze what becomes of the theorems in Example 11.19 when translated in $\tilde{T}$. There will be inconsistencies related to the fact that the Earth is sometimes viewed as fixed (when the Moon and the cannonballs are looked at), sometimes viewed as moving (when viewed as a planet). Try to fix these inconsistencies, for instance, by considering two different translations. There is a way out of these problems by replacing the purely kinematic theory T with a dynamical theory. (Kinematic means geometric i.e., not invoking forces whereas dynamical means involving forces.)

REMARK 43.7. Other physical theories can be given metamodels in a similar way. The logical content of this operation is never made explicit because it is too laborious, and often leads to contradictions if not done in minute detail. But it is important to understand how things work at least in principle. Problems such as these are not uncommon: set theory acts sometimes like a "straightjacket" for physical theories.

EXERCISE 43.8. (For those who took courses in physics)

1) Give a metamodel for Newton's dynamical theory of gravitation (based on forces).

2) Give a metamodel for classical mechanics of particles traveling in a field.

3) Give a metamodel for relativistic mechanics of particles moving freely.

4) Give a metamodel for a simple situation in quantum theory.

Next we consider the example of Peano arithmetic. The Peano axiom is the sentence in L_{set} that there exists a Peano triple. But if one tries to write this axiom in a language that only contains, say, a constant 0, binary functional symbols $+, \times$, and unary functional symbol s, together with equality $=$, then one is bound to fail because there is no way to say, in this language, that "for any subset something happens." Nevertheless there is a Peano system of axioms in such a language which we are going to consider below, together with a metamodel of the associated theory.

EXAMPLE 43.9. Consider the language L_{ar} with constant 0, binary functional symbols $+, \times$, and unary functional symbol s, together with equality $=$ and the standard connectives and separators. This is called the language of Peano arithmetic. We consider the following specific axioms:

A_1) $\forall x(s(x) \neq 0)$
A_2) $\forall x((x \neq 0) \rightarrow \exists y(s(y) = x)))$
A_3) $\forall x(x + 0 = x)$
A_4) $\forall x \forall y(x + (s(y)) = s(x + y))$
A_5) $\forall x(x \times 0 = 0)$
A_6) $\forall x \forall y(x \times s(y) = (x \times y) + x)$
A_P) $\forall w((P(0, w) \wedge \forall v(P(v, w) \rightarrow P(s(v), w))) \rightarrow \forall x(P(x, w)))$

where P runs through all formulas in L_{ar}^f with the appropriate number of free variables. The axioms A_P stand for a weak form of induction.

EXERCISE 43.10. Let $D(x)$ be the formula "$x \in \mathbb{N} \cup \{0\}$" in L_{set}^f. Let T_{ar} be the theory in L_{ar} with specific axioms $A_1, ..., A_6, A_P$; we call this theory the Peano arithmetic theory. Show that the theory T_{ar} has a metamodel with domain D in which $\tilde{L}, \tilde{T}$ coincide with L_{set}, T_{set}.

Here is another example of the use of metamodels which involves a variant of Grothendieck's universes; this is one of the ways out of the logical difficulties of category theory, to be discussed later.

EXAMPLE 43.11. Fix a set $\mathcal{U}$ (i.e., a constant in T_{set}) referred to in what follows as a universe and introduce a predicate still denoted by $\mathcal{U}$ equipped with the definition

$$\forall x(\mathcal{U}(x) \leftrightarrow (x \in \mathcal{U})).$$

Let $\tilde{L}$ be obtained from L_{set} by adding the above predicate. Also let $\tilde{\ }$ be the natural $\mathcal{U}$-translation of formulas from L_{set} into $\tilde{L}$. Finally let $\tilde{T}$ be the theory in the language $\tilde{L}$ generated by the axioms in ZFC, the collection $ZFC^{\tilde{\ }}$ of their $\mathcal{U}$-translations, and all sentences of the form $\mathcal{U}(c)$ where c is a constant in L_{set}. The above define a metamodel for T_{set} with domain $\mathcal{U}$. (So we are dealing here with a $\mathcal{U}$-translation of the language of set theory essentially *into itself* which is not the "identity!" See the exercise below. We will see a similar, but also very different, procedure when we discuss models of set theory in the chapter on models.) Intuitively $\tilde{T}$ is obtained from the set theory T_{set} by adding axioms saying that "all operations with sets that belong to $\mathcal{U}$, allowed by ZFC, lead to sets that belong to $\mathcal{U}$" and axioms saying that "all named sets in set theory are in $\mathcal{U}$." Of course the consistency of $\tilde{T}$ is even more problematic than the consistency of set theory T_{set}.

EXERCISE 43.12. Write down the axioms in $ZFC^{\tilde{\ }}$. Hint: If S is the singleton axiom

$$\forall x \exists y((x \in y) \wedge (\forall z((z \in y) \rightarrow (z = x))))$$

then its $\mathcal{U}$-translation $\tilde{S}$ is:

$$\forall x((x \in \mathcal{U}) \rightarrow (\exists y((y \in \mathcal{U}) \wedge (x \in y) \wedge (\forall z((z \in \mathcal{U}) \wedge (z \in y)) \rightarrow (z = x))))))).$$

EXERCISE 43.13. Prove (in $\tilde{T}$) that $\mathcal{U} \notin \mathcal{U}$.

Hint: Use the same arguments as in the Exercise 13.31 related to the "Russell paradox." Assume $\mathcal{U} \in \mathcal{U}$. Consider the set $S = \{x \in \mathcal{U}; x \notin x\}$ (use the separation axiom A in T_{set}.) Since $\mathcal{U} \in \mathcal{U}$, by the image $\tilde{A}$ in $ZFC^{\tilde{\ }}$ of the separation axiom A we get that $S \in \mathcal{U}$. Then analyze each of the cases $S \in S$ and $S \notin S$ and get, in each case, a contradiction.

CHAPTER 44

Categories

Categories are one of the most important unifying concepts of mathematics. In particular they allow to create bridges between various parts of mathematics. Categories were introduced by Eilenberg and MacLane partly motivated by work in homological algebra. A further input in the development of the concept was given by work of Grothendieck in algebraic geometry. Here we will only explore the definition of categories and we give some examples. We begin with the simpler concept of correspondence.

DEFINITION 44.1. A correspondence on a set $X^{(0)}$ is a triple $(X^{(1)}, \sigma, \tau)$ where $\sigma, \tau : X^{(1)} \to X^{(0)}$ are maps called source and target.

EXAMPLE 44.2. If $R \subset A \times A$ is a relation and $\sigma : R \to A$ and $\tau : R \to A$ are defined by $\sigma(a, b) = a$, $\tau(a, b) = b$ then (R, σ, τ) is a correspondence.

DEFINITION 44.3. Assume $(X^{(1)}, \sigma, \tau)$ is a correspondence on $X = X^{(0)}$. Then one can define sets

$$X^{(2)} \quad = \quad \{(a, b) \in X^2; \tau(b) = \sigma(a)\}$$

$$X^{(3)} \quad = \quad \{(a, b, c) \in X^3; \tau(c) = \sigma(b), \tau(b) = \sigma(a)\}, \quad \text{etc.}$$

We have natural maps $p_1, p_2 : X^{(1)} \to X^{(0)}$, $p_1(a, b) = a$, $p_2(a, b) = b$.

DEFINITION 44.4. A category is a tuple $\mathcal{C} = (X^{(0)}, X^{(1)}, \sigma, \tau, \mu, \epsilon)$ where $X^{(0)}$ is a set, $(X^{(1)}, \sigma, \tau)$ is a correspondence on $X^{(0)}$, and $\mu : X^{(2)} \to X^{(1)}$, $\epsilon : X^{(0)} \to X^{(1)}$ are maps. We assume that $\sigma \circ \epsilon = \tau \circ \epsilon = I$, the identity of $X^{(0)}$. Also we assume that the following diagrams are commutative:

$$
\begin{array}{ccc}
X^{(2)} \overset{\mu}{\to} X^{(1)} & X^{(2)} \overset{\mu}{\to} X^{(1)} & X^{(3)} \overset{\mu \times 1}{\to} X^{(2)} \\
p_1 \downarrow \quad \downarrow \tau \quad , & p_2 \downarrow \quad \downarrow \sigma \quad , & 1 \times \mu \downarrow \quad \downarrow \mu \quad . \\
X^{(1)} \overset{\tau}{\to} X^{(0)} & X^{(1)} \overset{\sigma}{\to} X^{(0)} & X^{(2)} \overset{\mu}{\to} X^{(0)}
\end{array}
$$

Finally we assume that the compositions

$$X^{(1)} \xrightarrow{1 \times (\epsilon \circ \sigma)} X^{(2)} \overset{\mu}{\to} X^{(1)}$$

and

$$X^{(1)} \xrightarrow{(\epsilon \circ \tau) \times 1} X^{(2)} \overset{\mu}{\to} X^{(1)}$$

are the identity of $X^{(1)}$, where

$$(1 \times (\epsilon \circ \sigma))(a) = (a, \epsilon(\sigma(a))),$$

$$((\epsilon \circ \tau) \times 1)(a) = (\epsilon(\tau(a)), a).$$

The set $X^{(0)}$ is called the set of objects of the category and is also denoted by $Ob(\mathcal{C})$. The set $X^{(1)}$ is called the set of arrows or morphisms and is sometimes denoted by $Mor(\mathcal{C})$. The map μ is called composition and we write $\mu(a,b) = a \star b$. The maps σ and τ are called the source and the target map, respectively. The map ϵ is called the identity. We set $\epsilon(x) = 1_x$ for all x. The first commutative diagram says that the target of $a \star b$ is the target of a. The second diagram says that the source of $a \star b$ is the source of b. In the third diagram (called associativity diagram) the map $\mu \times 1$ is defined as $(a,b,c) \mapsto (a \star b, c)$ while the map $1 \times \mu$ is defined by $(a,b,c) \mapsto (a, b \star c)$; the diagram then says that $(a \star b) \star c = a \star (b \star c)$. For $x,y \in X^{(0)}$ one denotes by $Hom(x,y)$ the set of all morphisms $a \in X^{(1)}$ with $\sigma(a) = x$ and $\tau(a) = y$. Instead of $a \in Hom(x,y)$ we also write $a : x \to y$. We say $a \in Hom(x,y)$ is an isomorphism if there exists $a' \in Hom(y,x)$ such that $a \star a' = 1_y$ and $a' \star a = 1_x$. (Then a' is unique and is denoted by a^{-1}.) A category is called a groupoid if all morphisms are isomorphisms.

What we called a category in the above definition is sometimes called a small category; since our presentation will involve universes (see below) we do not need to make any distinction between small categories and categories.

REMARK 44.5. Given a category
$$\mathcal{C} = (X^{(0)}, X^{(1)}, \sigma, \tau, \mu, \epsilon)$$
one can define the opposite category
$$\mathcal{C}^\circ = (X^{(0)}, X^{(1)}, \tau, \sigma, \mu \circ S, \epsilon)$$
where $S(a,b) = (b,a)$.

REMARK 44.6. Given a category
$$\mathcal{C} = (X^{(0)}, X^{(1)}, \sigma, \tau, \mu, \epsilon)$$
and a subset $Y \subset X^{(0)}$ one can define the full subcategory associated to Y by
$$\mathcal{C}_Y = (Y^{(0)}, Y^{(1)}, \tau_Y, \sigma_Y, \mu_Y, \epsilon_Y)$$
where $Y^{(0)} = Y$, $Y^{(1)} = \{a \in X^{(0)}; \sigma(a) \in Y, \tau(a) \in Y\}$, and $\tau_Y, \sigma_Y, \mu_Y, \epsilon_Y$ defined as the restrictions of $\tau, \sigma, \mu, \epsilon$, respectively.

In what follows we give some basic examples of categories. Some of the examples involve universes. For those examples we let $\mathcal{U}$ be a fixed universe and fix the metamodel $\tilde{L}, \tilde{T}, \tilde{\ }$ of set theory T_{set} attached to $\mathcal{U}$; cf. Example 43.11. (So $\tilde{T}$ is obtained from T_{set} by adding the corresponding axioms related to the universe.) For the various examples below that involve universes the correctness of the definitions depends on certain claims (typically that certain sets are maps, etc.) Those claims cannot be proved a priori in set theory T_{set} but they are trivially proved in $\tilde{T}$; this shows that the correctness of the definition of some of the categories below requires the axioms of $\tilde{T}$ hence *more axioms* than ZFC! The question of consistency of $\tilde{T}$ becomes then acute: one has to *believe* that $\tilde{T}$ is consistent in order to believe that the categories below are all well defined.

EXAMPLE 44.7. Define the category of sets (in a given universe $\mathcal{U}$) denoted by
$$\{\text{sets}\}$$
as follows:
$$X^{(0)} = \mathcal{U},$$
$$X^{(1)} = \{(A, B, F) \in \mathcal{U}^3; F \in B^A\},$$

$$\sigma(A, B, F) = A, \quad \tau(A, B, F) = B.$$

The following is a (trivially proved) theorem in $\tilde{T}$:

(*) *The set*

$$\mu = \{(((B, C, F), (A, B, G)), (A, C, H)) \in X^{(2)} \times X^{(1)}; H = F \circ G\}$$

is a map $X^{(2)} \to X^{(1)}$.

(see the Exercise below). Similarly we may define the map $\epsilon : X^{(0)} \to X^{(1)}$, $\epsilon(A) = I_A$ (identity of A). With above definitions we get a category. In the examples below we will encounter from time to time the same kind of phenomenon (where the correctness of definitions depends on theorems in $\tilde{T}$); we will not repeat the corresponding discussion but we will simply add everywhere the words "in a given universe."

EXERCISE 44.8. Prove that the above sentence (*) is a theorem in $\tilde{T}$. Hint: Use the fact that the axioms in $\tilde{T}$ involving $\mathcal{U}$ imply that if $F, G \in \mathcal{U}$ then $F \circ G \in \mathcal{U}$.

EXAMPLE 44.9. If in the example above we insist that all Fs are bijections we get a category called

$$\{\text{sets} + \text{bijections}\}.$$

This category is a groupoid.

EXAMPLE 44.10. Let A be a set and consider the category denoted by

$$\{\text{bijections of } A\}$$

defined as follows. We let $X^{(0)} = \{x\}$ be a set with one element, we let $X^{(1)}$ be the set of all bijections $F : A \to A$, we let σ and τ be the constant map $F \mapsto x$, we let μ be defined again by $\mu(F, G) = F \circ G$ (compositions of functions), and we let $\epsilon(F) = I_A$. This category is a groupoid.

EXAMPLE 44.11. Define the category

$$\{\text{ordered sets}\}$$

as follows. We take $X^{(0)}$ the set of all ordered sets $(A, \leq)$ with A in a given universe, we take $X^{(1)}$ to be the set of all triples $((A, \leq), (A', \leq'), F)$ with $(A, \leq), (A', \leq') \in X^{(0)}$ and $F : A \to A'$ increasing, we take μ to be again, composition, and we take $\epsilon(A, \leq) = I_A$.

EXAMPLE 44.12. Let $(A, \leq)$ be an ordered set. Define the category

$$\{(A, \leq)\}$$

as follows. We let $X^{(0)} = A$, we let $X^{(1)}$ be the set $\leq$ viewed as a subset of $A \times A$, we let $\sigma(a, b) = a$, $\tau(a, b) = b$, we let $\mu((a, b), (b, c)) = (a, c)$, and we let $\epsilon(a) = (a, a)$.

EXAMPLE 44.13. Equivalence relations give rise to groupoids. Indeed let A be a set with an equivalence relation $R \subset A \times A$ on it which we refer to as $\sim$. Define the category

$$\{(A, \sim)\}$$

as follows. We let $X^{(0)} = A$, we let $X^{(1)} = R$, we let $\sigma(a, b) = a$, $\tau(a, b) = b$, we let $\mu((a, b), (b, c)) = (a, c)$, and we let $\epsilon(a) = (a, a)$. This category is a groupoid.

EXAMPLE 44.14. We fix the type of algebraic structures below. (For instance we may fix two binary operations, one unary operation, and two given elements.) Define the category

$$\{\text{algebraic structures}\}$$

as follows. $X^{(0)}$ is the set of all algebraic structures $(A, \star, ..., \neg, ..., 1, ...)$ of the given type with A in a given universe, $X^{(1)}$ is the set of all triples

$$((A, \star, ..., \neg, ..., 1, ...), (A', \star', ..., \neg', ..., 1', ...), F)$$

with F a homomorphism, σ and τ are the usual source and target, and ϵ is the usual identity.

EXAMPLE 44.15. Here is a variant of the above example. Consider the category of rings

$$\{\text{commutative unital rings}\}$$

as follows. The set of objects $X^{(0)}$ is the set of all commutative unital rings $(A, +\times, 0, 1)$ (usually referred to as A) in a given universe and the set of arrows $X^{(1)}$ is the set of all triples (A, B, F) where A, B are rings and $F : A \to B$ is a ring homomorphism. Also the target, source, and identity are the obvious ones; the composition map is the usual composition.

EXAMPLE 44.16. Define the category

$$\{\text{topological spaces}\}$$

as follows. We let $X^{(0)}$ be the set of all topological spaces X in a given universe; we take $X^{(1)}$ to be the set of all triples (X, X', F) with $F : X \to X'$ continuous, we let μ be given by usual composition of maps, σ and τ the usual source and target maps, and ϵ the usual identity.

EXAMPLE 44.17. Here is a variation on the previous example. Define the category

$$\{\text{pointed topological spaces}\}$$

as follows. We let $X^{(0)}$ be the set of all pairs (X, x) where X is a topological space in a given universe and $x \in X$; we take $X^{(1)}$ to be the set of all triples $((X, x), (X', x'), F)$ with $F : X \to X'$ continuous, and $F(x) = x'$; we let μ be given by usual composition of maps, σ and τ the usual source and target maps, and ϵ the usual identity.

EXAMPLE 44.18. Define the category of groups

$$\{\text{groups}\}$$

defined as follows. The set $X^{(0)}$ of objects is the set of all groups whose set is in a given universe, and the set $X^{(1)}$ is the set of all the triples consisting of two groups G, H and a homomorphism between them. The source, target, and identity are the obvious ones.

EXAMPLE 44.19. If in the above example we restrict ourselves to groups which are Abelian we get the category of Abelian groups

$$\{\text{Abelian groups}\}.$$

EXAMPLE 44.20. A fixed group can be viewed as a category as follows. Fix a group $(G, \star, ', e)$. Then one can consider the category

$$\{G\}$$

where $X^{(0)} = \{x\}$ is a set consisting of one element, $X^{(1)} = G$, $\mu(a, b) = a \star b$, the source and the target are the constant maps, and $\epsilon(x) = e$.

EXAMPLE 44.21. Define the category

$$\{\text{vector spaces}\}$$

as follows. The set of objects $X^{(0)}$ is the set of all vector spaces, in a given universe, over a fixed field; the set $X^{(1)}$ of morphisms consists of all triples (V, W, F) with $F : V \to W$ a linear map; source, target, and identity are defined in the obvious way.

EXAMPLE 44.22. Define the category

$$\{\text{complex affine algebraic varieties}\}$$

as follows. The objects of the category (called complex affine algebraic varieties) are pairs $(\mathbb{C}^n, X)$ where X is a subset $X \subset \mathbb{C}^n$ for which there exist polynomials $f_1, ..., f_m \in \mathbb{C}[x_1, ..., x_n]$ such that

$$X = \{(a_1, ..., a_n) \in \mathbb{C}^n ; f_1(a_1, ..., a_n) = ... = f_m(a_1, ..., a_n) = 0\}.$$

(Lines, conics, and cubics introduced earlier are examples of complex affine algebraic varieties if one takes $R = \mathbb{C}$.) A morphism between $(\mathbb{C}^n, X)$ and $(\mathbb{C}^{n'}, X')$ is a map $F : X \to X'$ such that there exist polynomials $F_1, ..., F_{n'} \in \mathbb{C}[x_1, ..., x_n]$ with the property that for any $(a_1, ..., a_n) \in X$ we have

$$F(a_1, ..., a_n) = (F_1(a_1, ..., a_n), ..., F_{n'}(a_1, ..., a_n)).$$

Composition of morphisms is composition of maps.

EXAMPLE 44.23. Define the category

$$\{\text{ordinary differential equations}\}$$

as follows. The set of objects consists of all pairs $(\mathbb{R}^n, V)$ where $V : C^\infty(\mathbb{R}^n) \to C^\infty(\mathbb{R}^n)$ is a derivation which is $\mathbb{R}$-linear. (Such a V is called a vector field on $\mathbb{R}^n$.) A morphism $(\mathbb{R}^n, V) \to (\mathbb{R}^m, W)$ is a map $u : \mathbb{R}^n \to \mathbb{R}^m$ with smooth components such that for all $f \in C^\infty(\mathbb{R}^m)$ we have $V(f \circ u) = W(f) \circ u$. The reason why this is viewed as the right category for the theory of ordinary differential equations will be seen later.

EXERCISE 44.24. Check that, in all examples above, the axioms in the definition of a category are satisfied. (This is long and tedious but straightforward.)

CHAPTER 45

Functors

Each category describes, in some sense, the paradigm for one area of mathematics. The bridges between the various areas are realized by functors, cf. the definition and examples below.

DEFINITION 45.1. A functor $\Phi : \mathcal{C} \to \tilde{\mathcal{C}}$ between two categories

$$\mathcal{C} = (X^{(0)}, X^{(1)}, \sigma, \tau, \mu, \epsilon), \quad \text{and} \quad \tilde{\mathcal{C}} = (\tilde{X}^{(0)}, \tilde{X}^{(1)}, \tilde{\sigma}, \tilde{\tau}, \tilde{\mu}, \tilde{\epsilon})$$

is a pair of maps $(\Phi^{(0)}, \Phi^{(1)})$,

$$\Phi^{(0)} : X^{(0)} \to \tilde{X}^{(0)}, \quad \Phi^{(1)} : X^{(1)} \to \tilde{X}^{(1)}$$

such that the following diagrams are commutative:

$$
\begin{array}{ccc}
X^{(1)} & \xrightarrow{\Phi^{(1)}} & \tilde{X}^{(1)} \\
\sigma \downarrow & & \downarrow \tilde{\sigma} \\
X^{(0)} & \xrightarrow{\Phi^{(0)}} & \tilde{X}^{(0)}
\end{array}
\,,\quad
\begin{array}{ccc}
X^{(1)} & \xrightarrow{\Phi^{(1)}} & \tilde{X}^{(1)} \\
\tau \downarrow & & \downarrow \tilde{\tau} \\
X^{(0)} & \xrightarrow{\Phi^{(0)}} & \tilde{X}^{(0)}
\end{array}
\,,\quad
\begin{array}{ccc}
X^{(1)} & \xrightarrow{\Phi^{(1)}} & \tilde{X}^{(1)} \\
\epsilon \uparrow & & \uparrow \tilde{\epsilon} \\
X^{(0)} & \xrightarrow{\Phi^{(0)}} & \tilde{X}^{(0)}
\end{array}
$$

$$
\begin{array}{ccc}
X^{(2)} & \xrightarrow{\Phi^{(2)}} & \tilde{X}^{(2)} \\
\mu \downarrow & & \downarrow \tilde{\mu} \\
X^{(1)} & \xrightarrow{\Phi^{(1)}} & \tilde{X}^{(1)}
\end{array}
$$

where $\Phi^{(2)} : X^{(2)} \to \tilde{X}^{(2)}$ is the naturally induced map. One usually denotes both $\Phi^{(0)}$ and $\Phi^{(1)}$ by Φ. So compatibility with μ and $\tilde{\mu}$ reads

$$\Phi(a \star b) = \Phi(a) \star \Phi(b),$$

for all $(a, b) \in X^{(2)}$.

Here are a few examples of functors. We start with some "forgetful" functors whose effect is to "forget" part of the structure:

EXAMPLE 45.2. Consider the "forgetful" functor

$$\Phi : \{\text{commutative unital rings}\} \to \{\text{Abelian groups}\}$$

defined as follows. For $(R, +, \times, -, 0, 1)$ a commutative unital ring we let

$$\Phi(R, +, \times, -, 0, 1) = (R, +, -, 0),$$

which is an Abelian group. For any ring homomorphism F we set $\Phi(F) = F$, viewed as a group homomorphism.

EXAMPLE 45.3. Consider the "forgetful" functor

$$\Phi : \{\text{commutative unital rings}\} \to \{\text{Abelian groups}\}$$

defined as follows. For $(R, +, \times, -, 0, 1)$ a commutative unital ring we let

$$\Phi(R, +, \times, -, 0, 1) = (R^{\times}, \times, (\ )^{-1}, 1),$$

where $R^\times$ is the group of invertible elements of R, which is an Abelian group, and x^{-1} is the inverse of x. For any ring homomorphism F we let $\Phi(F)$ be the restriction of F to the invertible elements.

EXAMPLE 45.4. Consider the functor

$$\Phi : \{\text{Abelian groups}\} \to \{\text{groups}\}$$

defined as follows. For $(G, \star, ', e)$ an Abelian group we let

$$\Phi(G, \star, ', e) = (G, \star, ', e).$$

For any group homomorphism F we let $\Phi(F) = F$.

EXAMPLE 45.5. Consider the "forgetful" functor

$$\Phi : \{\text{groups}\} \to \{\text{sets}\}$$

defined as follows. For $(G, \star, ', e)$ a group we let

$$\Phi(G, \star, ', e) = G.$$

For any group homomorphism F we let $\Phi(F) = F$, as a map of sets.

EXAMPLE 45.6. Consider the following functor that is the prototype for some important functors in areas of mathematics called functional analysis and algebraic geometry. The functor is

$$\Phi : \{\text{topological spaces}\} \to \{\text{commutative unital rings}\}^\circ,$$

it takes values in the opposite of the category of commutative unital rings, and is defined as follows. For X a topological space we let

$$\Phi(X) = (C^0(X), +, \cdot, -, 0, 1),$$

where the latter is the following ring. The set $C^0(X)$ is the set of all continuous functions $f : X \to \mathbb{R}$, the addition $+$ and multiplication $\cdot$ are the pointwise operations, and $0(x) = 0,\ \ 1(x) = 1$. For any continuous map $F : X \to X'$ we let $\Phi(F) = F^*$ where, for $f : X' \to \mathbb{R}$, $F^*(f) = f \circ F$.

EXERCISE 45.7. Prove that all Φs in the examples above are functors. (This is tedious but straightforward.)

EXAMPLE 45.8. Consider the following functor that, again, is the prototype for some important functors in functional analysis and algebraic geometry. The functor is

$$\Phi : \{\text{commutative unital rings}\}^\circ \to \{\text{topological spaces}\},$$

and is defined as follows. Consider any commutative unital ring R. By an ideal in R we understand a subset $I \subset R$ such that I is a subgroup of R with respect with addition (i.e., $0 \in I$, $a + b \in I$, and $-a \in I$ for all $a, b \in I$), and $ab \in I$ for all $a \in R$ and $b \in I$. An ideal P in R is called a prime ideal if $P \neq R$ and whenever $ab \in P$ with $a \in R$ and $b \in R$ it follows that either $a \in P$ or $b \in P$. We let $Spec\ R$ be the set of all prime ideals in R. For any ideal I in R we let $D(I) \subset Spec\ R$ be the set of all prime ideals P such that $I \not\subset P$. Then the collection of all subsets of the form $D(I) \in \mathcal{P}(Spec\ R)$ is a topology on $Spec\ R$ called the Zariski topology. With this topology $Spec\ R$ becomes a topological space and we define

$$\Phi(R) = Spec\ R.$$

If $F : R \to R'$ is a ring homomorphism we define $\Phi(F) = F^* : Spec\ R' \to Spec\ R$ by $F^*(P') = F^{-1}(P)$.

EXERCISE 45.9. Prove that the collection $\{D(I); I$ an ideal in R$\}$ is a topology on $Spec\ R$. Prove that F^* is continuous. Prove that Φ is a functor.

EXERCISE 45.10. Prove that the ideals of $\mathbb{Z}$ are exactly the subgroups of $\mathbb{Z}$; hence they are of the form $\langle n \rangle$ for $n \in \mathbb{Z}$, $n \geq 0$. Prove that $\langle n \rangle$ is a prime ideal if and only if n is prime or $n = 0$. Prove that $Spec\ \mathbb{Z}$ is not a Hausdorff space.

EXERCISE 45.11. Consider the following functor that plays a key role in algebraic geometry. The functor is

$$\Phi : \{\text{complex affine algebraic varieties}\} \to \{\text{topological spaces}\},$$

and is defined as follows. If $(\mathbb{C}^n, X)$ is a complex affine algebraic variety, $X \subset \mathbb{C}^n$, then one can give X the topology induced from the Euclidean topology of $\mathbb{C}^n$ which we call the Euclidean topology on X; then X becomes a Hausdorff topological space and we let $\Phi(\mathbb{C}^n, X) = X$, with the Euclidean topology. For any morphism of complex affine algebraic varieties $F : X \to X'$ we let $\Phi(F) = F$ (which is continuous for the Euclidean topologies).

EXERCISE 45.12. Check that if $F : X \to X'$ is a morphism of complex affine algebraic varieties then F is continuous for the Euclidean topologies.

EXAMPLE 45.13. Consider the following functor that plays a key role in an area of mathematics called algebraic topology. The functor is

$$\Phi : \{\text{pointed topological spaces}\} \to \{\text{groups}\}$$

and is defined as follows. For (X, x) a pointed topological space we let

$$\Phi(X, x) = (\pi_1(X, x), \star, ', e),$$

where the latter is the following group (called the fundamental group of (X, x)). To define the set $\pi_1(X, x)$ we first define the set $\Pi(X, x)$ of all continuous maps $\gamma : [0, 1] \to X$ such that $\gamma(0) = \gamma(1) = x$; the elements γ are called loops. Next one defines a relation $\sim$ on $\Pi(X, x)$ called homotopy: two loops $\gamma_0, \gamma_1 : [0, 1] \to X$ are called homotopic (and write $\gamma_0 \sim \gamma_1$) if there exists a continuous map $F : [0, 1] \times [0, 1] \to X$ such that $F(t, 0) = \gamma_0(t)$, $F(t, 1) = \gamma_1(t)$, $F(0, s) = x$, $F(1, s) = x$, for all $t, s \in [0, 1]$. One proves that $\sim$ is an equivalence relation on $\Pi(X, x)$ and one defines the set $\pi_1(X, x)$ as the set of equivalence classes:

$$\pi_1(X, x) = \Pi(X, x)/ \sim .$$

The class of a loop γ is denoted by $[\gamma] \in \pi_1(X, x)$. On the other hand there is a natural "composition map"

$$\Pi(X, x) \times \Pi(X, x) \to \Pi(X, x), \quad (\gamma_1, \gamma_2) \mapsto \gamma_1 \star \gamma_2,$$

defined by $(\gamma_1 \star \gamma_2)(t) = \gamma_1(2t)$ for $0 \leq t \leq 1/2$ and $(\gamma_1 \star \gamma_2)(t) = \gamma_2(2t - 1)$ for $1/2 \leq t \leq 1$. (Note that $\gamma_1 \star (\gamma_2 \star \gamma_3) \neq (\gamma_1 \star \gamma_2) \star \gamma_3$ in general.) However one can prove that

(45.1) $$\gamma_1 \star (\gamma_2 \star \gamma_3) \sim (\gamma_1 \star \gamma_2) \star \gamma_3.$$

Define a binary operation on $\pi_1(X, x)$ by

$$[\gamma_1] \star [\gamma_2] = [\gamma_1 \star \gamma_2].$$

This makes $\pi_1(X, x)$ a group with identity $e = [\gamma_x]$ where $\gamma_x(t) = x$ for all t; associativity follows from Equation 45.1. For any continuous map $F : X \to X'$ we let $\Phi(F) = F_*$ where, for $\gamma \in \Pi(X, x)$, we let $F_*([\gamma]) = [F \circ \gamma]$.

EXERCISE 45.14. Prove that $\sim$ is an equivalence relation on $\Pi(X, x)$. Prove the homotopy in Equation 45.1. Show that the operation $\star$ is well defined on $\pi_1(X, x)$ and gives a group structure on $\pi_1(X, x)$. Check that the data above define a functor.

CHAPTER 46

Objectives

One can ask if there is a way to summarize the main objectives of modern mathematics. We would like this summary to transcend the particularities of the various fields in which the questions are being raised. The language of categories seems to be well adapted for this purpose, as we shall see in the examples below.

EXAMPLE 46.1. (Equations and solutions) Let $\mathcal{C}$ be a category as above. Let us define an equation to be a morphism $b \in Hom(y, z)$ and let $a \in Hom(x, z)$. Let us define the set of solutions in a of the equation b as the set

$$Sol(a, b) = \{c \in Hom(x, y); a = b \star c\}.$$

A large part of mathematics is devoted to "finding the set of solutions of given equations" in the sense above. Algebraic equations and differential equations can be put, for instance, into this setting.

In order to put algebraic equations into the framework above let us consider a simple situation in which $\mathcal{C}$ is the dual of the category of commutative unital rings. Let $f \in A = \mathbb{Z}[x_1, ..., x_n]$ be a polynomial in variables $x_1, ..., x_n$ with coefficients in $\mathbb{Z}$ and define an equivalence relation $\equiv_f$ on A by declaring that $u \equiv_f v$ for $u, v \in A$ if and only if there exists $w \in A$ such that $u - v = fw$. Write $A/(f)$ for the set $A/\equiv_f$ of equivalence classes. Then $A/(f)$ becomes a commutative unital ring with operations induced by the operations on A. Now let $b \in Hom(A/(f), \mathbb{Z})$ be the morphism corresponding to the natural ring homomorphism $\mathbb{Z} \to A/(f)$ and let $a \in Hom(R, \mathbb{Z})$ be the morphism corresponding to the natural homomorphism $\mathbb{Z} \to R$ where R is a field. Then we claim that there is a natural bijection

$$\psi : Sol(a, b) \to \{(c_1, ..., c_n) \in R^n; \ f(c_1, ..., c_n) = 0\}$$

given as follows: for any solution $c \in Hom(R, A/(f))$ corresponding to a ring homomorphism $c : A/(f) \to R$ we can attach the tuple $\psi(c) = (c([x_1]), ..., c([x_n]))$ where $[x_i] \in A/(f)$ is the equivalence class of x_i. This shows that the concept of solution in the category $\mathcal{C}$ corresponds to the usual concept of solution of an algebraic equation.

In order to put differential equations into a categorical framework start with a differential equation of the form

(46.1) $$\frac{d^r F}{dx^r} = Q(x, F(x), \frac{dF}{dx}(x), ..., \frac{d^{r-1}F}{dx^{r-1}}(x))$$

where $F \in C^\infty(\mathbb{R})$ and $Q \in C^\infty(\mathbb{R}^{r+1})$. Let now $\mathcal{C}$ be the category of ordinary differential equations. Consider the object $X = Z = (\mathbb{R}^1, D)$ with $DF = F'$ the usual derivative. Consider also the object $Y = (\mathbb{R}^{r+1}, V_Q)$ where for $f \in C^\infty(\mathbb{R}^{r+1})$,

$f = f(x, y_0, ..., y_{r-1})$, we set

$$(46.2) \ V_Q(f) = \frac{\partial f}{\partial x} + y_1 \frac{\partial f}{\partial y_0} + y_2 \frac{\partial f}{\partial y_1} + ... + y_{r-1} \frac{\partial f}{\partial y_{r-2}} + Q(x, y_0, ..., y_{r-1}) \frac{\partial f}{\partial y_{r-1}}.$$

We let $a : X \to X = Z$ be the identity and $b : Y \to X$ be defined by the first projection $\mathbb{R}^{r+1} \to \mathbb{R}$, $(x, y_0, ..., y_{r-1}) \mapsto x$. Then there is a natural bijection

$$\Psi : Sol(a, b) \to \{F \in C^\infty(\mathbb{R}); \ F \text{ is a solution to } 46.1\}$$

given as follows. For any solution $c \in Sol(a, b)$ given by a map $c : \mathbb{R} \to \mathbb{R}^{r+1}$, $c(x) = (x, c_0(x), c_1(x), ..., c_{r-1}(x))$, we let $\Psi(c) = F$ with $F(x) = c_0(x)$. This shows that the concept of solution in the category $\mathcal{C}$ corresponds to the usual concept of solution of a differential equation of the form 46.1.

EXERCISE 46.2. In the notation above define the operations on $A/(f)$ and check that $A/(f)$ is a ring. Prove that ψ is well defined and a bijection.

EXERCISE 46.3. In the notation above prove that Ψ is well defined and a bijection.

EXAMPLE 46.4. (Symmetries) Let $\mathcal{C}$ be a category and $x \in Ob(\mathcal{C})$. We denote by $Aut(x)$ the set of isomorphisms in $Hom(x, x)$. Then $(Aut(x), \star, (\ )^{-1}, 1_x)$ is a group; it is referred to as the automorphism group of x and is viewed as the "group of symmetries of x." Many problems of modern mathematics boil down to computing this group.

Here is an example. Let $\mathcal{C}$ be the category of commutative unital rings. Let $f \in \mathbb{Q}[x]$ be a polynomial in one variable and let $Z = \{\alpha_1, ..., \alpha_n\}$ be the set of all roots of f in $\mathbb{C}$. Let

$$K_f = \{P(\alpha_1, ..., \alpha_n); \ P \in \mathbb{Q}[x_1, ..., x_n]\} \in Ob(\mathcal{C}).$$

(This is actually a field, called the splitting field of f.) Then $Aut(K_f)$ is the group of all ring isomorphisms $g : K_f \to K_f$; this group is called the Galois group of f over $\mathbb{Q}$ and is sometimes denoted by G_f; it plays a central role in number theory. For $g \in G_f$ we have that $g(Z) = Z$ so g induces a permutation σ_g of $\{1, ..., n\}$ such that $g(\alpha_i) = \alpha_{\sigma_g(i)}$ for all i. We get an injective homomorphism $G_f \to S_n$, $g \mapsto \sigma_g$.

Here is another example. Let $\mathcal{C}$ be the category of ordinary differential equations. For the object $(\mathbb{R}^{r+1}, V_Q)$ with V_Q as in 46.2, the group $Aut(\mathbb{R}^{r+1}, V_Q)$ is viewed as the group of symmetries of the Equation 46.1. For $V = 0$ the group $Aut(\mathbb{R}^n, V)$ is called the diffeomorphism group of $\mathbb{R}^n$ and is usually denoted by $Diff(\mathbb{R}^n)$.

Here is another example. Let $\mathcal{C}$ be the category of all vector spaces over a field R. Then there is a natural isomorphism $Aut(R^n) \to GL_n(R)$.

EXERCISE 46.5. Look at other examples of categories and analyze the Aut groups of their objects.

EXAMPLE 46.6. (Classification problem and invariants) Let $\mathcal{C}$ be a category. Then there is an equivalence relation $\simeq$ on $Ob(\mathcal{C})$ defined as follows: for $x, y \in Ob(\mathcal{C})$ we have $x \simeq y$ if and only if there exists an isomorphism $x \to y$. One can consider the set of equivalence classes

$$Ob(\mathcal{C})/\simeq .$$

"Describing" this set is referred to as the classification problem for the objects of category $\mathcal{C}$ and many important problems in modern mathematics boil down to the classification problem for an appropriate category.

For some categories this is trivial; for instance if $\mathcal{C}$ is the full subcategory of the category of vector spaces over a field R consisting of the finite dimensional vector spaces then there is a natural bijection

$$Ob(\mathcal{C})/\simeq \ \to \ \mathbb{N} \cup \{0\}, \ [V] \mapsto \dim \ V.$$

However for other categories $\mathcal{C}$ such as the category of topological spaces or the category of complex affine algebraic varieties no description is available for $\mathcal{C}/\simeq$; partial results (e.g., results for full subcategories of these categories or variants of these categories) are known and some are very deep.

Another way to formulate (or weaken) the problem is via systems of invariants for objects of a category $\mathcal{C}$. If S is a set a system of invariants for $\mathcal{C}$ is a map

$$I : Ob(\mathcal{C}) \to S$$

such that for any $x, y \in Ob(\mathcal{C})$ with $x \simeq y$ we have $I(x) = I(y)$. Any system of invariants defines a map

$$\overline{I} : Ob(\mathcal{C})/\simeq \ \to \ S, \ \overline{I}([x]) = I(x).$$

If the latter is an injection we say I is a complete system of invariants. So the classification problem is the same as the problem of (explicitly) finding a complete system of invariants I and finding "all possible invariants" (i.e., the image of I). For instance if $\mathcal{C}$ is the category of finite dimensional vector spaces over a field R then $I = \dim$ is a complete system of invariants and I is surjective. A weaker problem for a general category is to find an "interesting" (not necessarily complete) system of invariants.

Some important theorems in mathematics claim the equality of a priori unrelated invariants $I' : Ob(\mathcal{C}) \to S$ and $I'' : Ob(\mathcal{C}) \to S$. As corollaries one sometimes obtains interesting equalities between (integer or real) numbers.

Finally note that functors can produce systems of invariants as follows. If $\Phi : \mathcal{C} \to \mathcal{C}'$ is a functor then we have a natural map

$$I : Ob(\mathcal{C})/\simeq \ \to \ Ob(\mathcal{C}')/\simeq'.$$

So any system of invariants for $\mathcal{C}'$ induces a system of invariants for $\mathcal{C}$. This is one of the main ideas of algebraic topology (respectively algebraic geometry) for which $\mathcal{C}$ is the category of topological spaces (respectively complex affine algebraic varieties or variants of it) and $\mathcal{C}'$ is the category of groups, rings, vector spaces, etc.

REMARK 46.7. It is far from being the case that all main questions of mathematics have a structural flavor that fits into the categorical framework explained above. What *is* the case, however, is that one expects that behind many of the important results of mathematics there is a more general, structural "explanation" that *does* fit into the categorical viewpoint; looking for such "explanations" is a modern trend in mathematics called *categorification*.

Part 3

Mathematical logic

CHAPTER 47

Models

We briefly indicate here how one can create a "mirror" of pre-mathematical logic within mathematics (indeed within algebra). What results is a subject called mathematical logic (also referred to as formal logic). A special place in mathematical logic is played by models/model theory which we shall explain below in some detail. In particular we will discuss models of ZFC itself (and hence of mathematics itself); this concept of model is entirely different from that of metamodels in the chapter on metamodels. In the discussion of models of ZFC we will introduce what one can call formalized set theory. We will discuss models of other (simpler) theories as well such as the theory of Peano arithmetic.

The mirror of pre-mathematical logic in mathematics is not entirely accurate: there is no one to one correspondence between pre-mathematical logic and mathematical logic. But as a general principle if a concept (say *crocodile*) is present in pre-mathematical logic its mirror in mathematical logic will acquire the adjective *formal* in front (e.g., will be called a *formal crocodile*). The dichotomy non-formal/formal indicates the difference between set theory in pre-mathematical logic and what we shall call formalized set theory in mathematical logic. There is another dichotomy, the syntactic/semantic dichotomy; *semantic* means, roughly, coming from translation of sentences whereas *syntactic* means coming from the shape of the sentences.

All definitions below are in set theory.

DEFINITION 47.1. Define the sets

$$\begin{aligned} T &= \{c, f_1, f_2, f_3 ..., r_1, r_2, r_3, ...\} \\ V &= \{x_1, x_2, x_3, ...\} \\ W &= \{\vee, \wedge, \neg, \forall, \exists, =, (,)\}. \end{aligned}$$

V is called the set of variables; W is called the set of logical symbols. We sometimes write $x, y, z, ...$ instead of $x_1, x_2, x_3, ...$ By a T-partitioned set we mean in what follows a set S together with a map $S \to T$. We let $S_t \subset S$ the preimage of $t \in T$. Let S be a T-partitioned set; the elements of S_c are called constant symbols; the elements of S_{f_n} are called n-ary function symbols; the elements of S_{r_n} are called n-ary relation symbols. For any such S we consider the set

$$\Lambda_S = V \cup W \cup S$$

(referred to as the formal language attached to S).Then one considers the set of words $\Lambda_S^\star$ with letters in Λ_S. One defines (in an obvious way, imitating the metadefinitions in the chapters on "pre-mathematical" logic) what it means for an element $\varphi \in \Lambda_S^\star$ to be an S-formula or an S-formula without free variables (the latter are referred to as sentences). One denotes by $\Lambda_S^f \subset \Lambda_S^*$ the set of all S-formulas and by $\Lambda_S^s \subset \Lambda_S^f$ the set of all S-sentences.

REMARK 47.2. Needless to say in the above $c, f_i, r_i, x_i, \wedge, \dots$ are sets. One can take, if one wants, $c = 0$, $f_i = (1, i)$, $r_i = (2, i)$, $x_i = (3, i)$, $\wedge = 1$, $\vee = 2, \dots$, but that would be confusing. Our notation also has a confusing side. For instance the symbol $\wedge$ has two different uses: it is, of course, a connective in L_{set}; but in 47.1 it is also a set so it is a constant in L_{set}. Similarly $x, y, z, \dots$ are the variables in L_{set}; but in 47.1 they are sets so they are constants in L_{set}. One needs to keep in mind this double use of the various symbols.

EXAMPLE 47.3. Assume $\rho \in S_{r_3}$, $a \in S_c$, and $x, z \in V$. Define the word

$$\varphi = \forall x \exists z (\rho(x, z, a)) = (\forall, x, \exists, z, (, \rho, (, x, z, a,),),)).$$

Words are sets so φ is a set. Also φ is an S-formula. But the word $\exists a \forall z (\rho(x, z, a))$ is not an S-formula because constants cannot have quantifiers $\forall, \exists$ in front of them. Also the word $\forall x \exists z (\rho(x, a))$ is not an S-formula because ρ is "supposed to have 3 arguments." If $\square \in S_{f_2}$ and a, x, z are as above then the word

$$\varphi = \forall x \exists z (\square(z, a) = z)$$

is an S-formula.

REMARK 47.4. One can define binary operations $\wedge^*$ and $\vee^*$ on Λ_S^s and a unary operation $\neg^*$ on Λ_S^s. For instance, $\wedge^* : \Lambda_S^s \times \Lambda_S^s \to \Lambda_S^s$ is defined by $\varphi \wedge^* \psi = (\varphi) \wedge (\psi)$. We have $\varphi \wedge^* (\psi \wedge^* \eta) = (\varphi) \wedge ((\psi) \wedge (\eta)) \neq (\varphi \wedge^* \psi) \wedge^* \eta$. So Λ_S^s is not a Boolean algebra with respect to these operations. For simplicity we continue to write $\wedge, \vee, \neg$ in place of $\wedge^*, \vee^*, \neg^*$.

EXAMPLE 47.5. Let M be a set. We let $S_c(M) = M$. For $n \in \mathbb{N}$ we set $S_{r_n}(M) = \mathcal{P}(M^n)$, the set of n-ary relations on M and $S_{f_n}(M) \subset \mathcal{P}(M^{n+1})$ the set of maps $M^n \to M$. We consider the T-partitioned set $S(M)$, union of the above. Then we can consider the formal language $\Lambda_{S(M)}$. An assignment in M is a map

$$\mu : V = \{x_1, x_2, x_3, \dots\} \to M.$$

For any assignment μ there exists a unique map $v_{M,\mu} : \Lambda_{S(M)}^f \to \{0, 1\}$ which is a homomorphism with respect to $\vee, \wedge, \neg$, is compatible (in an obvious sense) with $\forall, \exists$, and satisfies obvious properties with respect to relational and functional symbols, and also with μ. If φ has no free variables we write $v_M(\varphi)$ in place of $v_{M,\mu}(\varphi)$. Here is an example of the "obvious properties" referred to above:

$$\forall x \forall y (((x \in M) \wedge (y \in \mathcal{P}(M))) \to ((v_M(x \in y) = 1) \leftrightarrow (x \in y)))$$

which translates into argot as follows: the value of v_M on the formula $x \in y$ is 1 if and only if $x \in y$. Similar properties are satisfied by v_M for more complicated formulae.

Next we discuss "semantics" of formal languages. The word "semantics" here is taken in a metaphorical sense.

DEFINITION 47.6. By a translation of Λ_S into $\Lambda_{S'}$ we understand a map

$$\mathfrak{m} : S \to S'$$

which is compatible with the partitions (in the sense that $\mathfrak{m}(S_t) \subset S'_t$ for all $t \in T$). For any $\varphi \in \Lambda_S^f$ one can form, in an obvious way, a formula $\mathfrak{m}(\varphi) \in \Lambda_{S'}^f$ obtained from φ by replacing the constants and relational and functional symbols by their images under $\mathfrak{m}$, respectively. So we get a map

$$\mathfrak{m} : \Lambda^f_S \to \Lambda^f_{S'}$$

which is a homomorphism with respect to $\vee, \wedge, \neg$, compatible with $\forall, \exists$.

An S-structure (or simply a structure if S is understood) is a pair $\mathcal{M} = (M, \mathfrak{m})$ where M is a set and $\mathfrak{m}$ is a translation

$$\mathfrak{m} : S \to S(M).$$

So we get a map

$$\mathfrak{m} : \Lambda^f_S \to \Lambda^f_{S(M)}$$

which is a homomorphism with respect to $\vee, \wedge, \neg$, compatible with $\forall, \exists$. Fix an assignment μ and set $v_{M,\mu} = v_M$. We have a natural map

$$v_M : \Lambda^f_{S(M)} \to \{0, 1\}$$

which is again a homomorphism compatible with $\forall, \exists$. So we may consider the composition

$$v_{\mathcal{M}} = v_M \circ \mathfrak{m} : \Lambda^f_S \to \{0, 1\}.$$

We say that a sentence $\varphi \in \Lambda^s_S$ is satisfied in the structure $\mathcal{M}$ if $v_{\mathcal{M}}(\varphi) = 1$. This concept is independent of μ. This is a variant of Tarski's *semantic definition of truth*: one can define in set theory the predicate *is true in* $\mathcal{M}$ by the definition:

$$\forall x((x \text{ is true in } \mathcal{M}) \leftrightarrow ((x \in \Lambda^s_S) \wedge (v_{\mathcal{M}}(x) = 1))).$$

(We will continue to NOT use the word *true* in what follows, though.) If $v_{\mathcal{M}}(\varphi) = 1$ we also say that $\mathcal{M}$ is a model of φ and we write $\mathcal{M} \models \varphi$. (So "is a model" and "$\models$" are predicates that are being added to L_{set}.) We say a set $\Phi \subset \Lambda^s_S$ of sentences is satisfied in a structure (write $\mathcal{M} \models \Phi$) if all the formulas in Φ are satisfied in this structure. We say that a sentence $\varphi \in \Lambda^s_S$ is a semantic formal consequence of a set of sentences $\Phi \subset \Lambda^s_S$ (and we write $\Phi \models \varphi$) if φ is satisfied in any structure in which Φ is satisfied. Here the word semantic is used because we are using translations; and the word formal is being used because we are in mathematical logic rather than in pre-mathematical logic. A sentence $\varphi \in \Lambda^s_S$ is valid (or a formal tautology) if it is satisfied in any structure, i.e., if $\emptyset \models \varphi$. We say a sentence φ is satisfiable if there is a structure in which it is satisfied. Two sentences φ and ψ are semantically formally equivalent if each of them is a semantic formal consequence of the other, i.e $\varphi \models \psi$ and $\psi \models \varphi$; write $\varphi \approx \psi$. Note that the quotient $\Lambda^s_S / \approx$ is a Boolean algebra in a natural way. Moreover each structure $\mathcal{M}$ defines a homomorphism $v_{\mathcal{M}} : \Lambda^s_S / \approx \to \{0, 1\}$ of Boolean algebras.

EXAMPLE 47.7. Algebraic structures can be viewed as models. Here is an example. Let

$$S = \{\star, \iota, e\}$$

with e a constant symbol, $\star$ a binary function symbol, and ι a unary function symbol. Let Φ_{gr} be the set of S-formulas $\Phi_{gr} = \{\varphi_1, \varphi_2, \varphi_3\}$ where

$$
\begin{aligned}
\varphi_1 &= \forall x \forall y \forall z (x \star (y \star z) = (x \star y) \star z) \\
\varphi_2 &= \forall x (x \star e = e \star x = x) \\
\varphi_3 &= \forall x (x \star \iota(x) = \iota(x) \star x = x).
\end{aligned}
$$

Then a group is simply a model of Φ_{gr} above. We also say that Φ_{gr} is a set of axioms for the formalized theory of groups.

More generally:

DEFINITION 47.8. A formalized (or formal) system of axioms is a pair (S, Φ) where S is a T-partitioned set and Φ is a subset of Λ_S^s; then define

$$\Theta = \Phi^{\vDash} = \{\varphi \in \Lambda_S^s; \Phi \vDash \varphi\};$$

Θ is called the formal theory generated by the system of axioms (S, Φ).

REMARK 47.9. One thinks of $\Phi^{\vDash}$ as semantically defined because its definition involves models and hence translations. One may define a "syntactic" version of this set namely the set

$$\Phi^{\vdash} = \{\varphi \in \Lambda_S^s; \Phi \vdash \varphi\};$$

where $\vdash$ is the predicate added to L_{set} meaning "φ provable from Φ" in an obvious sense that imitates the definition of proof in pre-mathematical logic. We will not make this precise here, neither will we use the following theorem of Gödel which is usually referred to as Gödel's completeness theorem. This theorem intuitively says that syntactic and semantic provability coincide.

THEOREM 47.10. $\Phi^{\vdash} = \Phi^{\vDash}$.

EXERCISE 47.11. Write down a set of symbols S and a set of formulas Φ which is a set of axioms for the formalized theory of:
   1) commutative unital rings;
   2) fields;
   3) ordered sets;
   4) vector spaces over a given field;
   5) Boolean algebras;
   6) sets.
Can one find such a pair S, Φ in the following cases?
   7) well ordered sets;
   8) topological spaces;
   9) the ring of integers.

CHAPTER 48

Incompleteness

Our aim here is to state and explain Gödel's "incompleteness theorems of set theory and arithmetic."

DEFINITION 48.1. Formalized (or formal) set theory is the formal theory Θ_{set} generated by (S_{set}, Φ_{set}) where $S_{set} = \{\in\}$ and Φ_{set} consists of the ZFC axioms of set theory viewed as elements of $\Lambda_{set} = \Lambda_{\{\in\}}$.

REMARK 48.2. Unlike the language L_{set} of set theory, which has constants (e.g., the witnesses), Λ_{set} above has no constants! Also, of course, L_{set} is a text (more precisely a collection of symbols) whereas Λ_{set} is a set (hence, again, a text, but consisting of one symbol only). Similarly the theory T_{set} is a text (more precisely a collection of strings of symbols) while the formal theory Θ_{set} is itself a set (hence, again, a text consisting of one symbol only). However any sentence P in L_{set}^s without constants gives rise in an obvious way to a sentence (which we still denote by P) in Λ_{set}^s and vice versa. We will use this abuse of notation freely in what follows. For instance the sentence $\exists x(x \in b)$ in L_{set}^s is a string of symbols in L_{set} consisting of a quantifier $\exists$, a variable x, a separator (, again the variable x, a relational predicate $\in$, a constant b, and another separator). The sentence $\exists x(x \in b)$ in L_{set}^s gives rise to a S_{set}-sentence,

$$\varphi = \exists x(x \in b) = (\exists, x, (, x, \in, b,)) \in \Lambda_{set}^s.$$

Note that φ is a word, hence a set, hence a constant in L_{set}. So, using appropriate definitions, we attached to a string consisting of a quantifier, a variable, a separator, etc., in L_{set} a single constant in L_{set}.

METADEFINITION 48.3. We let T'_{set} the theory generated by T_{set} and all sentences in L_{set}^s of the form

$$P \leftrightarrow (P \in \Theta_{set})$$

where P in the left hand side is viewed in L_{set}^s and P in the right hand side is viewed in Λ_{set}^s.

REMARK 48.4. By Theorem 47.10 the new axioms above have the form

(48.1) $$P \leftrightarrow (P \in \Phi_{set}^\vdash).$$

Both the left hand side and the right hand side of the sentences 48.1 have a syntactic nature; but they are very different in that the left hand involves P as a text (more precisely a string of symbols) whereas the right hand side involves P as a set (hence as a symbol). In this way these new axioms "postulate a short-circuit" between texts and sets. This is not the standard way to present Gödel's theorems that follows because it is not standard to distinguish between texts and sets; since we chose to make this distinction we are forced to add these new axioms to T_{set}. The translation in English of these new axioms 48.1 is something like:

"P if and only if there exists a proof for P."

The latter is, of course, very misleading because the words in English are ambiguous.

DEFINITION 48.5. A formal theory Θ generated by a system of axioms (S, Φ) is called formally inconsistent if there exists a sentence $\varphi \in \Lambda_S^s$ such that $\varphi \in \Theta$ and $\neg\varphi \in \Theta$. Θ is called formally consistent if it is not formally inconsistent. Θ is called formally complete if for any sentence $\varphi \in \Lambda_S^s$ either $\varphi \in \Theta$ or $\neg\varphi \in \Theta$. Θ is called formally incomplete if it is not formally complete.

Let P_{con} be the sentence in L_{set}^s (in set theory) expressing (in an "obvious way" which will not be made precise here) the formal consistency of formal set theory Λ_{set}. Then one can consider P_{con} as a sentence in Λ_{set}^s and then one can consider "$P_{con} \notin \Theta_{set}$" as a sentence in L_{set}. Gödel proved that this is a theorem in T_{set}':

THEOREM 48.6. $P_{con} \notin \Theta_{set}$.

With regards to the continuum hypothesis we have the following results due to Gödel and Cohen, respectively. Let P_{CH} be the sentence in L_{set}^s expressing the continuum hypothesis, and view P_{CH} as a sentence in Λ_{set}^s. We have the following theorems in T_{set}':

THEOREM 48.7. $\neg P_{CH} \notin \Theta_{set}$.

THEOREM 48.8. $P_{CH} \notin \Theta_{set}$.

COROLLARY 48.9. Θ_{set} is formally incomplete.

Next we want to analyze the integers in mathematical logic.

REMARK 48.10. The Peano axiom in set theory is the sentence that there exists a Peano triple. One can view this as a sentence in $\Lambda_{set} = \Lambda_{\{\in\}}$ of formalized set theory; indeed the words "there exists a set z such that for any subset y ..." can be written as
$$\exists z \forall y ((\forall x (x \in y) \to (x \in z)) \to ...).$$
But if one tries to write this axiom in the formal language $\Lambda_{ar} = \Lambda_{\{+,\times,s,0\}}$ where $+, \times$ are binary functions, s is a unary function, and 0 is a constant then one is bound to fail because there is no way to say, in this formal language, that for any subset something happens. (If one says $\forall x$ then in any S-structure $(M, \mathfrak{m})$ the variable x is translated as an element of M and not as a subset of M.) Nevertheless there is a formal Peano system of axioms in the formal language Λ_{ar} which we are going to consider below.

DEFINITION 48.11. Consider the set $S = S_{ar} = \{+, \times, s, 0\}$ where $+, \times$ are binary functional symbols, s is a unary functional symbol, and 0 is a constant. Let Φ_{ar} be the following sentences in $\Lambda_{ar} = \Lambda_{S_{ar}}$ (called the formal Peano axioms): $A_1, ..., A_6, A_P$ (obtained from the Peano axioms 43.9). So this is a countable set of axioms; the latter series of axioms stands for a weak form of induction. The formal system of axioms of Peano arithmetic is the pair (S_{ar}, Φ_{ar}). The theory Θ_{ar} generated by (S_{ar}, Φ_{ar}) is called the Peano arithmetic.

Here is Gödel's incompleteness theorem of arithmetic. It is a theorem in T_{set}'.

THEOREM 48.12. Θ_{ar} is formally incomplete.

Sketch of proof. We follow the informal presentation in the first pages of Gödel's original article, in combination with facts proved later in that paper. We refer to loc. cit. for the details. Also for any sentence P in L_{ar}^s we continue to denote by P its image in Λ_{ar}^s and also its translation in L_{set}^s and the image of that translation in Λ_{set}^s.

The first step is to "tag" formulae by integers. One knows there is a bijection between the set of symbols Λ_{set} and $\mathbb{N}$; fix such a bijection and, for convenience, write this informally as

$$\Lambda_{set} = \{s_1, s_2, s_3, ...\}.$$

Let $p : \mathbb{N} \to \mathbb{N}$ be the function defined by letting $p(i)$ be the ith prime. Then define the "Gödel numbering function"

$$G : \Lambda_{set}^* \to \mathbb{N}$$

by the rule

$$G(s_{i_1} s_{i_2} s_{i_3} ...) = p(1)^{p(i_1)} p(2)^{p(i_2)} p(3)^{p(i_3)} ....$$

Clearly G is injective and "definable in L_{ar}" in the sense that its graph is given by an appropriate formula in L_{ar}^f in two variables. Let $\Sigma = G(\Lambda_{set}^f)$ and Σ' be the image by G of the set of all formulae with exactly one free variable. Let $F : \Sigma \to \Lambda_{set}^f$ be the inverse of $G : \Lambda_{set}^f \to \Sigma$.

The second step is to prove that the predicate "belongs to Θ_{ar}" is "definable" in L_{ar} in the sense that there is a formula B in L_{ar}^f with one free variable such that for all $n \in \Sigma$,

$$(F(n) \in \Theta_{ar}) \leftrightarrow B(n).$$

This is the heart of the proof: it amounts to showing that "provability of a sentence is an arithmetic property of the tag of the sentence." This is very laborious but not exceedingly hard.

The third step is to show that there is a formula S in L_{ar}^f in two variables defining the following "substitution" function $S : \Sigma' \times \mathbb{N} \to \mathbb{N}$: $S(a, b) = c$ if and only if $F(c)$ is the formula obtained from the formula $F(a)$ by replacing the unique free variable of $F(a)$ by b; in other words $F(S(a, b)) = F(a)(b)$. Now for x a variable we consider the formula P in L_{ar}^f having the property that, viewed as a formula in Λ_{ar}^f, it satisfies (together with B and S viewed as sets)

$$P = \neg B(S(x, x)) \in \Lambda_{ar}^f \quad \text{(equality of sets)}.$$

Since P has exactly one free variable there exists $m \in \Sigma$ such that $P = F(m)$. The key sentence to be considered now is

$$P(m) = \neg(B(S(m, m))) \in \Lambda_{ar}^s.$$

Note that

$$F(S(m, m)) = F(m)(m) = P(m),$$

hence

$$S(m, m) = G(P(m)),$$

hence

(48.2) $\neg P(m) = \neg\neg B(G(P(m))) \leftrightarrow B(G(P(m))) \leftrightarrow (P(m) \in \Theta_{ar}) \to P(m).$

The last step of the proof is to check the following:

(48.3) $P(m) \notin \Theta_{ar};$

(48.4) $\neg P(m) \notin \Theta_{ar}.$

To check 48.3 assume $P(m) \in \Theta_{ar}$ and seek a contradiction. By 48.2 we get both $P(m)$ and $\neg P(m)$ which is a contradiction.

To check 48.4 assume $\neg P(m) \in \Theta_{ar}$ and seek a contradiction. Since

$$(\neg P(m) \in \Theta_{ar}) \to (\neg P(m))$$

we get $\neg P(m)$. By 48.2 we get $P(m)$. This is a contradiction.

$\square$

REMARK 48.13. Since Theorems 48.6, 48.7, 48.8 are sentences in L^s_{set} they have, according to the conventions in the present course, no meaning and no truth value. In particular any interpretation of these theorems as saying something about mathematics itself transcends the paradigm of the present course.

Bibliography

[1] Cantor, G. 1955. *Contributions to the Founding of the Theory of Transfinite Numbers.* New York: Dover.

[2] Chomsky, N. 2006. *Language and Mind.* Cambridge University Press.

[3] Gödel, K. 1992. *On formally undecidable propositions of Principia Mathematica and related systems.* New York: Dover.

[4] Hilbert, D. 1994. *Foundations of Geometry.* La Salle: Open Court.

[5] Hilbert, D., Bernays, P. 1934. *Grundlagen der Mathematik.* Berlin.

[6] Kant, I. 1991. *Critique of Pure Reason.* London: J. M. Dent & Sons.

[7] Manin, Yu. I. 2009. *A Course in Mathematical Logic for Mathematicians.* New York: Springer.

[8] Quine, W. V. 1986. *Philosophy of Logic.* Cambridge: Harvard University Press.

[9] Russell, B. 1993. *Introduction to Mathematical Philosophy.* New York: Dover.

[10] Tarski, A. 1995. *Introduction to Logic and to the Methodology of Deductive Sciences*, New York: Dover.

[11] Weyl, H. 1963. *Philosophy of Mathematics and Natural Sciences.* New York: Atheneum.

[12] Wittgenstein, L. 2001. *Tractatus Logico-Philosophicus.* Routledge: Taylor and Francis.

Index